STUDENT SOLUTIONS MANUAL

Elementary

JAY R. SCHAFFER

Statistics

PICTURING THE WORLD

LARSON ■ FARBER

PRENTICE HALL, Upper Saddle River, NJ 07458

Acquisitions Editor: Kathy Boothby Sestak
Supplement Editor: Joanne Wendelken
Special Projects Manager: Barbara A. Murray
Production Editor: Jonathan Boylan
Supplement Cover Manager: Paul Gourhan
Supplement Cover Designer: PM Workshop Inc.
Manufacturing Buyer: Alan Fischer

Printed in the United States of America

10 9 8 7 6 5 4 3 2

ISBN 0-13-040070-X

Prentice-Hall International (UK) Limited, London
Prentice-Hall of Australia Pty. Limited, Sydney
Prentice-Hall Canada, Inc., Toronto
Prentice-Hall Hispanoamericana, S.A., Mexico
Prentice-Hall of India Private Limited, New Delhi
Prentice-Hall (Singapore) Pte. Ltd.
Prentice-Hall of Japan, Inc., Tokyo
Editora Prentice-Hall do Brazil, Ltda., Rio de Janeiro

CONTENTS

Introduction to Statistics

1.1 AN OVERVIEW OF STATISTICS

1.1 Try It Yourself Solutions

1a. The population consists of the prices per gallon of regular gasoline at all gasoline stations in the U.S.

b. The sample consists of the prices per gallon of regular gasoline at the 800 surveyed stations.

c. The data set consists of the 800 prices.

2a. Because the numerical measure of $1,119,537,215 is based on the entire collection of player's salaries, it is from a population.

b. Because the numerical measure is a characteristic of a population, it is a parameter.

3a. Descriptive statistics involve the statement "76% of women and 60% of men had a physical examination within the previous year."

b. An inference drawn from the study is that a higher percentage of women had a physical examination within the previous year.

1.1 EXERCISE SOLUTIONS

1. A sample is a subset of a population.

3. False. A statistic is a measure that describes a sample characteristic.

5. True

7. The data set is a population since it is a collection of the ages of all the state governors.

9. The data set is a sample since the collection of the 500 students is a subset within the population of the university's 2000 students.

11. Population: Party of registered voters in Bucks County.

Sample: Party of Bucks County voters responding to phone survey.

13. Population: Ages of adult Americans who own computers.

Sample: Ages of adult Americans who own Dell computers.

15. Population: Collection of all infants.

Sample: Collection of the 33,043 infants in the study.

17. Population: Collection of all American women.

Sample: Collection of the 546 American women surveyed.

19. Statistic. the value $1,320,000 is a numerical description of a sample of one month out of a year.

21. Statistic. 10% is a numerical description of a sample of computer users.

23. The statement "56% are the primary investor in their household" is an application of descriptive statistics.

    An inference drawn from the sample is that an association exists between American women and being the primary investor in their household.

25. Answers vary.

1.2 DATA CLASSIFICATION

1.2 Try It Yourself Solutions

1a. One data set contains names of cities and the other contains city populations.

b. City: Nonnumerical
Population: Numerical

c. City: Qualitative
Population: Quantitative

2. (1a) The final standings represent a ranking of hockey teams.

   (1b) ordinal

   (2a) The collection of phone numbers represent labels. No mathematical computations can be made.

   (2b) nominal

3. (1a) The collection of body temperatures represent data that can be ordered, but makes no sense written as a ratio.

   (1b) Interval

   (2a) The collection of heart rates represent data that can be ordered and written as a ratio that makes sense.

   (2b) Ratio

1.2 EXERCISE SOLUTIONS

1. Nominal and ordinal

3. True

5. False. Data at the ordinal level are qualitative or quantitative.

7. Qualitative, because telephone numbers are merely labels.

9. Quantitative

11. Ordinal. Data can be arranged in order, but differences between data entries make no sense.

13. Ratio. Two data values can be formed so one data value can be expressed as a multiple of another. There can be twice as many fish of one species than of another.

15. Interval. The data can be ordered and you can calculate meaningful differences between data entries. It is not meaningful to say one temperature is twice as hot as another.

17. Ordinal

19. Nominal

21. Interval data can be ordered and differences between entries can be calculated. Ratio data has all the properties of interval data with the addition that a ratio of two data values can be formed so one data value can be expressed as a multiple of another.

1.3 EXPERIMENTAL DESIGN

1.3 Try It Yourself Solutions

1. (1a) Focus: Effect of exercise on senior citizens.
   Population: Collection of all senior citizens.

   (1b) Experiment

   (2a) Focus: Effect of radiation fallout on senior citizens.
   Population: Collection of all senior citizens.

   (2b) Sampling

2a. Example: start with the first digits 92630782 . . .

   b. 92|63|07|82|40|19|26

   c. 63, 7, 40, 19, 26

3. (1) Because the students were readily available in your class, this is convenience sampling

   (2) Because the students were ordered in a manner such that every 25th student is selected, this is systematic sampling.

1.3 EXERCISE SOLUTIONS

1. False. Using stratified sampling guarantees that member of each group within a population will be sampled.

3. False. To select a systematic sample, a population is ordered in some way and then members of the population are selected at regular intervals.

5. In this study, you want to measure the effect of a treatment (using a fat substitute) on the human digestive system. So, you would want to perform an experiment.

7. Because it is impractical to create this situation, you would want to use a simulation.

9. Each U.S. telephone number has a equal chance of being dialed, so this is a simple random sample. Telephone sampling only samples those individuals who have telephones, are available, and are willing to respond, so this is a possible source of bias.

11. Because the students were chosen due to their convenience of location (leaving the library), this is a convenience sample. There may be an association between time spent at the library and drinking habits, so this is a possible source of bias.

13. Because a random sample of out-patients were selected, this is a simple random sample.

15. Because a sample is taken from each one-acre subplot (stratum), this is a stratified sample.

17. Because every twentieth name on a list is being selected, this is a systematic sample.

19. Question is biased since it already suggests that drinking fruit juice is good for you. The question might be rewritten as "How does drinking fruit juice affect your health?"

21. The households sampled represent various locations, ethnic groups, and income brackets. Each of these variables is considered a stratum.

23. (a) Advantage: Allows respondent to express some depth and shades of meaning in the answer.
    Disadvantage: Not easily quantified and difficult to compare surveys.

    (b) Advantage: Easy to analyze results.
    Disadvantage: May not provide appropriate alternatives and may influence the opinion of the respondent.

CHAPTER 1 REVIEW EXERCISE SOLUTIONS

1. Population: Collection of all U.S. VCR owners.
   Sample: Collection of the 898 VCR owners that were sampled.

3. Population: Collection of all U.S. ATM's.
   Sample: Collection of 860 ATM's that were sampled.

5. The team payroll is a parameter since it is a numerical description of a population (entire baseball team) characteristic.

7. Since "89 students" is describing a characteristic of a population (University of Arizona students), it is a parameter.

9. Quantitative because monthly salaries are numerical measurements.

11. Quantitative because ages are numerical measurements.

13. Interval. It makes no sense saying that 100 degrees is twice as hot as 50 degrees.

15. Nominal. The data is categorical and cannot be arranged in a meaningful order.

17. Because judges keep accurate records of charitable donations, you could take a census.

19. In this study, you want to measure the effect of a treatment (plant hormone) on chrysanthemums. You would want to perform an experiment.

21. Because random telephone numbers were generated and called, this is a simple random sample.

23. Since each community is considered a cluster and every pregnant woman in a selected community is surveyed; this is a cluster sample.

25. Since grades are considered strata and 25 students are sampled from each stratum, this is a stratified sample.

27. Telephone sampling only samples individuals who have telephones, are available, and are willing to respond.

29. The selected communities may not be representative of the entire area.

CHAPTER 1 QUIZ SOLUTIONS

1. Population: Collection of all individuals with sleep disorders.
   Sample: Collection of 163 patients in study.

2. (a) Statistic. 53% is a characteristic of a sample of parents.

   (b) Parameter. 67% is a characteristic of the entire union (population).

3. (a) Qualitative, since student identification numbers are merely labels.

   (b) Quantitative, since a test score is a numerical measure.

4. (a) Nominal. Players may be ordered numerically, but there is no meaning in this order.

   (b) Ratio. It makes sense to say that the number of products sold during the 1st quarter was twice as any as sold in the 2nd quarter.

5. (a) In this study, you want to measure the effect of a treatment (low dietary intake if vitamin C and iron) on lead levels in adults. You want to perform an experiment.

   (b) Since it would be difficult to survey every individual within 500 miles of your home, sampling should be used.

6. (a) Because people were chosen due to their convenience of location (on the beach), this is a convenience sample.

   (b) Since every fifth part is selected from an assembly line, this is a systematic sample

7. (a) False. A statistic is a numerical measure that describes a sample characteristic.

   (b) False. Ratio data represents the highest level of measurement.

Descriptive Statistics

2.1 FREQUENCY DISTRIBUTIONS AND THEIR GRAPHS

2.1 Try It Yourself Solutions

1a. The number of classes (6) is stated in the problem.

b. Min = 0   Max = 63 Class width = $\dfrac{(63 - 0)}{6}$ = 10.5 = >11

c.

Lower limit	Upper limit
0	10
11	21
22	32
33	43
44	54
55	65

d. See part (e).

e.

Class	Frequency, f
0–10	27
11–21	13
22–32	16
33–43	7
44–54	11
55–65	3

2a. See part (b).

b.

Class	Frequency, f	Midpoint	Relative frequency	Cumulative frequency
0–10	27	5	0.3506	27
11–21	13	16	0.1688	40
22–32	16	27	0.2078	56
33–43	7	38	0.0909	63
44–54	11	49	0.1429	74
55–65	3	60	0.0390	77
	77		1	

c. Over 35% of the population is under 11 years old. Less than 4% of the population is over 54 years old.

3a.

Class Boundaries
−0.5–10.5
10.5–21.5
21.5–32.5
32.5–43.5
43.5–54.5
54.5–65.5

b. Use class midpoints for the horizontal scale and frequency for the vertical scale.

c.

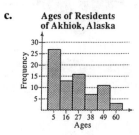

Ages of Residents of Akhiok, Alaska

d. Most of the residents are less than 32 years old.

4a. Use class midpoints for the horizontal scale and frequency for the vertical scale.

b. See part (c).

c.

Ages of Residents of Akhiok, Alaska

d. The population of Akhiok, Alaska is predominantly made up of young people.

5.

Ages of Residents of Akhiok, Alaska

6a. Use upper class boundaries for the horizontal scale and cumulative frequency for the vertical scale.

b. See part (c).

c.

Ages of Residents of Akhiok, Alaska

d. Approximately 63 residents are less than 45 years old.

7. See the solution to Try It Yourself problem 3.

2.1 EXERCISE SOLUTIONS

1. By organizing the data into a frequency distribution, patterns within the data may become more evident.

3. False. The midpoint frequency of a class is the frequency of the class divided by the sample size.

5. Class width = 10

Class	Frequency	Class Boundaries	Midpoint	Cumulative frequency
20–29	10	19.5–29.5	24.5	10
30–39	132	29.5–39.5	34.5	142
40–49	284	39.5–49.5	44.5	426
50–59	300	49.5–59.5	54.5	726
60–69	175	59.5–69.5	64.5	901
70–79	65	69.5–79.5	74.5	966
80–89	25	79.5–89.5	84.5	991

7. Least frequency ≈ 10
Greatest frequency ≈ 300
Class width = 10

9. (a) 50   (b) 12.5–13.5 lbs.   (c) 24   (d) 19.5 lbs.

11. (a) Class with greatest relative frequency: 8–9 in.
Class with least relative frequency: 17–18 in.

(b) Greatest relative frequency ≈ 0.195
Least relative frequency ≈ 0.005

(c) Approximately 0.015

13. Class with greatest frequency: 500–550
Class with least frequency: 250–300 or 700–750

15.

Class	Frequency	Midpoint	Relative frequency	Cumulative frequency
0–7	8	3.5	0.32	8
8–15	8	11.5	0.32	16
16–23	3	19.5	0.12	19
24–31	3	27.5	0.12	22
32–39	3	35.5	0.12	25
	25		1	

17.

Class	Frequency	Midpoint	Relative frequency	Cumulative frequency
1000–2019	12	1509.5	0.5455	12
2020–3039	3	2529.5	0.1364	15
3040–4059	2	3549.5	0.0909	17
4060–5079	3	4569.5	0.1364	20
5080–6099	1	5589.5	0.0455	21
6100–7119	1	6609.5	0.0455	22
	22		1	

July Sales for Representatives

Class with greatest frequency: 1000–2019
Class with least frequency: 5080–6099; 6100–7119

19.

Class	Frequency	Midpoint	Relative frequency	Cumulative frequency
291–318	4	304.5	0.1818	4
319–346	3	332.5	0.1364	7
347–374	2	360.5	0.0909	9
375–402	4	388.5	0.1818	13
403–430	3	416.5	0.1364	16
431–458	3	444.5	0.1364	19
459–486	1	472.5	0.0455	20
487–514	2	500.5	0.0909	22
	22		1	

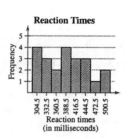

Class with greatest frequency: 291–318; 375–402
Class with least frequency: 459–486

21.

Class	Frequency	Midpoint	Relative frequency	Cumulative frequency
146–169	6	157.5	0.2308	6
170–193	9	181.5	0.3462	15
194–217	3	205.5	0.1154	18
218–241	6	229.5	0.2308	24
242–265	2	253.5	0.0769	26
	26		1	

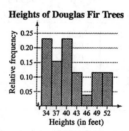

Class with greatest relative frequency: 170–193
Class with least relative frequency: 242–265

23.

Class	Frequency	Midpoint	Relative frequency	Cumulative frequency
33–35	6	34	0.2308	6
36–38	4	37	0.1538	10
39–41	6	40	0.2308	16
42–44	3	43	0.1154	19
45–47	1	46	0.0385	20
48–50	3	49	0.1154	23
51–53	3	52	0.1154	26
	26		1	

Class with greatest relative frequency: 33–35; 39–41
Class with least relative frequency: 45–47

25.

Class	Frequency	Relative frequency	Cumulative frequency
50–53	1	0.0417	1
54–57	0	0.0000	1
58–61	4	0.1667	5
62–65	9	0.3750	14
66–69	7	0.2917	21
70–73	3	0.1250	24
	24	1	

Location of the greatest increase in frequency: 62–65

27.

Class	Frequency	Relative frequency	Cumulative frequency
2–4	9	0.3214	9
5–7	6	0.2143	15
8–10	7	0.2500	22
11–13	3	0.1071	25
14–16	2	0.0714	27
17–19	1	0.0357	28
	28	1	

Location of the greatest increase in frequency: 2–4

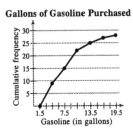

Gallons of Gasoline Purchased

29.

Class	Frequency	Midpoint	Relative frequency	Cumulative frequency
47–57	1	52	0.05	1
58–68	1	63	0.05	2
69–79	5	74	0.25	7
80–90	8	85	0.4	15
91–101	5	96	0.25	20
	20		1	

Class with greatest frequency: 80–90
Classes with least frequency: 47–57 and 58–68

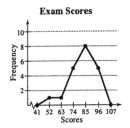

Exam Scores

31. (a)

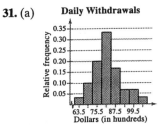

Daily Withdrawals

(b) $9,600 because only 10% of the time were the withdrawals larger.

(c) 16.7% of the time you would run out of cash. 16.7% of the withdrawals were larger than $9,000.

33.

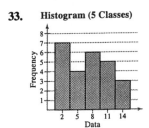

Histogram (5 Classes)

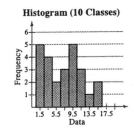

Histogram (10 Classes)

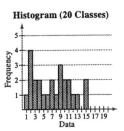

Histogram (20 Classes)

Given the three histograms (5, 10, and 20 classes), it appears that 5 classes would be best. The histograms with 10 and 20 classes have classes with zero frequencies in them. Not much is gained by using more than 5 classes.

2.2 MORE GRAPHS AND DISPLAYS

2.2 Try It Yourself Solutions

1a.
```
0|
1|
2|
3|
4|
5|
6|
```

b. Key: 3 | 5 = 35
```
0|6371750321869204165546824
1|72501703611026
2|8714725153968
3|9362403121
4|8756291
5|04302651
6|3
```

c. Key: 3 | 5 = 35
```
0|0111222334445555666677889
1|00011122356677
2|1123455677889
3|0112233469
4|1256789
5|00123456
6|3
```

d. It appears that the residents of Akhiok are a young population with most of the ages being below 40 years old.

2ab.
```
0|011122233444
0|555666677889
1|000111223
1|56677
2|11234
2|55677889
3|01122334
3|69
4|12
4|56789
5|001234
5|56
6|3
```

3a. Use ages for the horizontal axis.

b.

Ages of the Residents of Akhiok

c. It appears that a large percentage of the population is younger than 40 years old.

4a.

Transportation	Passengers Frequency	Relative frequency	Central angle
Bus	348	0.3460	124.6
Air	363.1	0.3610	130.0
Subway	274.6	0.2730	98.3
Amtrak	20.1	0.0200	7.2
	1005.8	1	

b.

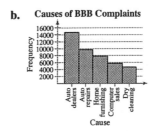

1985 Intercity Passenger Travel

c. It appears that subway and train travel is nearly the same but approximately 4% of the 1985 bus travelers are now taking airplanes.

5a.

Cause	Frequency
Auto Dealers	14668
Auto Repair	9728
Home Furnishing	7792
Computer Sales	5733
Dry Cleaning	4649

b.

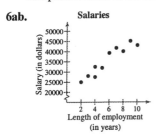

Causes of BBB Complaints

c. It appears that the auto industry (dealers and repair shops) account for the largest portion of complaints filed at the BBB.

6ab.

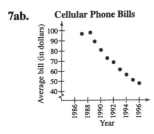

Salaries

c. It appears that the longer an employee is with the company, the larger his/her salary will be.

7ab.

Cellular Phone Bills

c. It appears that the average monthly bill for cellular telephone subscribers has decreased significantly over the last 10 years.

2.2 EXERCISE SOLUTIONS

1. Quantitative: Stem-and-Leaf Plot, Dot Plot, Histogram, Time Series Chart

Qualitative: Pie Chart, Pareto Chart

3. a

5. d

7. 27, 32, 41, 43, 43, 44, 47, 47, 48, 50, 51, 51, 52, 53, 53, 53, 54, 54, 54, 54, 55, 56, 56, 58, 59, 68, 68, 68, 73, 78, 78
Max: 78   Min: 27

9. 13, 13, 14, 14, 14, 15, 15, 15, 15, 15, 16, 17, 17, 18, 19
Max: 19   Min: 13

11. Time series chart since the data is taken at regular intervals over a period of time.

13. Stem-and-leaf plot or dot plot since the data can be used to illustrate the overall distribution of premium gasoline prices.

15. (a) Key: 2|4 = 24

```
1 | 9
2 | 11122233444566677777899
3 | 0
```

(b) Key: 2|4 = 24

```
1 | 9
2 | 11122233444
2 | 566677777899
3 | 0
```

It appears that using two rows for each stem displays the data better.

17. Key: 3|3 = 33

```
3 | 233459
4 | 01134556678
5 | 133
6 | 0069
```

Most elephants tend to drink less than 55 gallons of water per day.

19. Key: 17|5 = 175

```
16 | 48
17 | 113455679
18 | 13446669
19 | 0023356
20 | 18
```

It appears that most farmers charge 17 to 19 cents per pound of apples.

21.

Housefly Lifespans

It appears that the lifespan of a fly tends to be between 4 and 14 days.

23.

1995 NASA Budget

It appears that 40% of NASA's 1995 budget went to human space flight.

25.

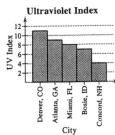

Ultraviolet Index

It appears that Denver, CO has nearly three times as much UV exposure than Concord, NH.

27.

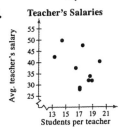

Teacher's Salaries

It appears that a teacher's average salary decreases as the number of students per teacher increases.

29.

Price of Whole Milk

It appears that the price of whole milk has increased over the past 7 years.

31.

Sales for Company A

When data is taken at regular intervals over a period of time, a time series chart should be used.

2.3   MEASURES OF CENTRAL TENDENCY

2.3 Try It Yourself Solutions

1a. $\Sigma x = 1745$

b. $\bar{x} = \Sigma x/n = 1745/77 = 22.66$

c. The typical age of a resident of Akhiok is 22.66 years old.

2a. 0, 1, 1, 1, 2, 2, 2, 3, 3, 4, 4, 4, 5, 5, 5, 6, 6, 6, 6, 7, 7, 8, 8, 9, 10, 10, 10, 11, 11, 11, 12, 12, 13, 15, 16, 16, 17, 17, 21, 21, 22, 23, 24, 25, 25, 26, 27, 27, 28, 28, 29, 30, 31, 31, 32, 32, 33, 33, 34, 36, 39, 41, 42, 45, 46, 47, 48, 49, 50, 50, 51, 52, 53, 54, 55, 56, 63

b. median = 21

c. Half of the residents of Akhiok are younger than 21 and half are older than 21.

3a. 0, 1, 1, 1, 2, 2, 2, 3, 3, 4, 4, 4, 5, 5, 5, 6, 6, 6, 7, 7, 8, 8, 9, 10, 10, 10, 11, 11, 12, 12, 13, 15, 16, 16, 17, 17, 21, 21, 22, 23, 24, 25, 25, 26, 27, 27, 28, 28, 29, 30, 31, 31, 32, 32, 33, 36, 39, 41, 42, 45, 46, 47, 48, 49, 50, 50, 51, 52, 53, 54, 55, 63

b. median $= (17 + 21)/2 = 19$

4a.

Class	Frequency	Age	Frequency	Age	Frequency
0	1	17	2	39	1
1	3	21	2	41	1
2	3	22	1	42	1
3	2	23	1	45	1
4	3	24	1	46	1
5	3	25	2	47	1
6	4	26	1	48	1
7	2	27	2	49	1
8	2	28	2	50	2
9	1	29	1	51	1
10	3	30	1	52	1
11	3	31	2	53	1
12	2	32	2	54	1
13	1	33	2	55	1
15	1	34	1	56	1
16	2	36	1	63	1

b. The age that occurs with the greatest frequency is 6 years old.

c. The mode of the ages of the residents of Akhiok is 6 years old.

5a. "Yes" occurs with the greatest frequency (169).

b. The mode of the responses to the survey is "Yes".

6a. $\bar{x} = \Sigma x/n = 410/19 \approx 21.58$

median $= 21$

mode $= 20$

b. The mean in example 6 ($\bar{x} = 23.75$) was heavily influenced by the age 65. Neither the median nor the mode was affected as much by the age 65.

7ab.

Source	Score x	Weight w	$x \cdot w$
Test Mean	86	0.50	43
Mid-Term	96	0.15	14.4
Final	98	0.20	19.6
Computer Lab	98	0.10	9.8
Homework	100	0.05	5
		1.00	91.8

c. $\bar{x} = \dfrac{\Sigma(x \cdot w)}{\Sigma w} = \dfrac{91.8}{1.00} = 91.8$

d. The weighted mean for the course is 91.8.

8abc.

Class	Midpoint x	Frequency f	$x \cdot f$
0-10	5	27	135
11-21	16	13	208
22-32	27	16	432
33-43	38	7	266
44-54	49	11	539
55-65	60	3	180
		77	1760

d. $\bar{x} = \dfrac{\Sigma(x \cdot f)}{n} = \dfrac{1760}{77} = 22.86$

e. The average age of a resident of Akhiok is approximately 22.86.

2.3 EXERCISE SOLUTIONS

1. False. The mean is the measure of central tendency most likely to be affected by an extreme value.

3. False. All quantitative data sets have a median.

5. Skewed right since the tail of the distribution extends to the right.

7. Uniform since the left and right halves of the distribution are approximately mirror images.

9. (7) since the distribution values range from 1 to 12 and have (approximately) equal frequencies.

11. (8) since the distribution has a maximum value of 90 and is skewed left due to a few students scoring much lower than the majority of the students.

13. (a) $\bar{x} = 6.23$
median = 6
mode = 5

(b) Median appears to be the best measure of central tendency since the distribution is skewed.

15. (a) $\bar{x} = 4.57$
median = 4.8
mode = 4.8

(b) Median appears to be the best measure of central tendency since there are outliers present.

17. (a) $\bar{x} = 97$
median = 97.2
mode = 94.8, 95.4, 97.2, 103.1

(b) Median appears to be the best measure of central tendency since the distribution is skewed.

19. (a) x = not possible
median = not possible
mode = "Worse"

(b) Mode appears to be the best measure of central tendency since the data is at the nominal level of measurement.

21. (a) $\bar{x} = 170.63$
median = 169.3
mode = not possible

(b) Mean appears to be the best measure of central tendency since there are no outliers.

23. (a) $\bar{x} = 22.6$
median = 19
mode = 14

(b) Median appears to be the best measure of central tendency since the distribution is skewed.

25. (a) $\bar{x} = 14.11$
median $= 14.25$
mode $= 2.5$

(b) Mean appears to be the best measure of central tendency since there are no outliers.

27. A = mode (data entry that occurred most often)
B = median (left of mean in skewed right dist.)
C = mean (right of median in skewed right dist.)

29.

Source	Score, x	Weight, w	x · w
Homework	85	0.15	12.75
Quiz	80	0.20	16
Quiz	92	0.20	18.4
Quiz	76	0.20	15.2
Final Exam	93	0.25	23.25
		1	85.6

$$\bar{x} = \frac{\Sigma(x \times w)}{\Sigma w} = \frac{85.6}{1} = 85.6$$

31.

Grade	Points, x	Credits, w	x · w
B	3	3	9
B	3	3	9
A	4	4	16
D	1	2	2
C	2	3	6
		15	42

$$\bar{x} = \frac{\Sigma(x \times w)}{\Sigma w} = \frac{42}{15} = 2.8$$

33.

Midpoint, x	Frequency, f	x · f
61	3	183
64	4	256
67	7	469
70	2	140
	16	1048

$$\bar{x} = \frac{\Sigma(x \times f)}{n} = \frac{1048}{16} \approx 65.5$$

35.

Midpoint, x	Frequency, f	x · f
4.5	57	256.5
14.5	68	986
24.5	36	882
34.5	55	1897.5
44.5	71	3159.5
54.5	44	2398
64.5	36	2322
74.5	14	1043
84.5	8	676
	389	13620.5

$$\bar{x} = \frac{\Sigma(x \times f)}{n} = \frac{13,620.5}{389} \approx 35.01$$

37.

Class	Midpoint, x	Frequency, f
3–4	3.5	3
5–6	4	8
7–8	7	4
9–10	2	2
11–12	11.5	2
13–14	13.5	1
		20

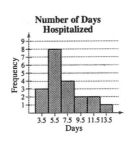

Shape: Positively skewed

39.

Class	Midpoint, x	Frequency, f
62–64	63	3
65–67	66	7
68–70	69	9
71–73	72	8
74–77	75	3
		30

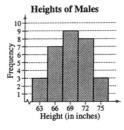

Shape: Symmetric

41. (a) $\bar{x} = 6.01$
median $= 6.01$

(b) $\bar{x} = 5.95$
median $= 6.01$

(c) mean

43. (a) Mean should be used since Car A has the highest mean of the three.

(b) Median should be used since Car B has the highest median of the three.

(c) Mode should be used since Car C has the highest mode of the three.

45. (a) $\bar{x} = 49.23$

(b) median $= 46.5$

(c)
```
1 | 1 3
2 | 2 8              median
3 | 6 6 6 7 7 8
4 | 1 3 4 6
5 | 1 1 1 3         mean
6 | 1 2 3 4
7 | 2 2 4 6
8 | 5
9 | 0
```

(d) Positively skewed

47. Two different symbols are needed since they describe a measure of central tendency for two different sets of data (sample is a subset of the population).

2.4   MEASURES OF VARIATION

2.4 Try It Yourself Solutions

1a. Min = 23 Max = 58

  b. Range = Max − Min = 58 − 23 = 35

  c. The range of the starting salaries for Corporation B is 35 or $35,000 (much larger than range of Corporation A).

2a. $\mu = \dfrac{\Sigma x}{N} = \dfrac{415}{10} = 41.5$

bc.

Salary, x	$x - u$
23	−18.5
29	−12.5
32	−9.5
40	−1.5
41	−0.5
41	−0.5
49	7.5
50	8.5
52	10.5
58	16.5
	0

3ab.

Salary, x	$x - u$	$(x - u)^2$
23	−18.5	342.25
29	−12.5	156.25
32	−9.5	90.25
40	−1.5	2.25
41	−0.5	0.25
41	−0.5	0.25
49	7.5	56.25
50	8.5	72.25
52	10.5	110.25
58	16.5	272.25
	0	1102.5

  c. $\sigma^2 = \dfrac{[\Sigma(x - \mu)^2]}{N} = \dfrac{1102.5}{10} = 110.25$

  d. $\sigma = \sqrt{\sigma^2} = \sqrt{110.25} = 10.5$

  e. The population variance is 110.25 and the population standard deviation is 10.5 or $10,500.

4a. See solution for 3ab.

  b. $s^2 = \dfrac{[\Sigma(x - \bar{x})^2]}{(n - 1)} = \dfrac{1102.5}{9} = 122.5$

  c. $s = \sqrt{s^2} = \sqrt{122.5} \approx 11.07$

5a. (Enter data in computer or calculator)

  b. $\bar{x} = 21.64,$ $s = 4.06$

6a. 7, 7, 7, 7, 7, 13, 13, 13, 13, 13

b. $\sigma = \sqrt{\dfrac{\Sigma(x - \mu)^2}{N}} = \sqrt{\dfrac{90}{10}} = 3$

7a. $64 - 61.25 = 2.75 = 1$ standard deviation

b. 34%

c. The estimated percent of the heights that are between 61.25 and 64 inches is 34%.

8a. $31.6 - 2(19.5) = -7.4 = >0$

b. $31.6 + 2(19.5) = 70.6$

c. At least 75% of the data lie within 2 standard deviations of the mean.

d. At least 75% of the population of Alaska is between 0 and 70.6 years old.

9a.

x	f	xf
0	10	0
1	19	19
2	7	14
3	7	21
4	5	20
5	1	5
6	1	6
	50	85

b. $\bar{x} = \dfrac{\Sigma xf}{n} = \dfrac{85}{50} = 1.7$

c.

$x - \bar{x}$	$(x - \bar{x})^2$	$(x - \bar{x})^2 \cdot f$
−1.70	2.8900	28.9
−0.70	0.4900	9.31
0.30	0.0900	0.63
1.30	1.6900	11.83
2.30	5.2900	26.45
3.30	10.9800	10.89
4.30	18.4900	18.49
		106.5

d. $s = \sqrt{\dfrac{[\Sigma(x - \bar{x})^2 f]}{(n - 1)}} = \sqrt{\dfrac{106.5}{49}} \approx 1.47$

10a.

Class	x	f	xf
0	0	33	0
1-99	50	4	200
100-199	150	10	1500
200-299	250	13	3250
300-399	350	12	4200
400-499	450	11	4950
500+	650	17	11050
		100	25150

b. $\bar{x} = \dfrac{\Sigma xf}{n} = \dfrac{25,150}{100} = 251.5$

c.

$x - \bar{x}$	$(x - \bar{x})^2$	$(x - \bar{x})^2 \cdot f$
−251.5	63252.25	2087324.25
−201.5	40602.25	162409
−101.5	10302.25	103022.5
−1.5	2.25	29.25
98.5	9702.25	116427
198.5	39402.25	433424.75
398.5	158802.25	2699638.25
		5602275

d. $s = \sqrt{\dfrac{[\Sigma(x - \bar{x})^2 f]}{(n - 1)}} = \sqrt{\dfrac{5,602,275}{99}} \approx 237.9$

2.4 EXERCISE SOLUTIONS

1. The range is the difference between the maximum and minimum values of a data set. The advantage of the range is that it is easy to calculate. The disadvantage is that it uses only two entries from the data set.

3. Range = max − min = 96 − 23 = 73

5. A deviation, $(x - \mu)$, is the difference between an observation, x, and the mean of the data, μ. The sum of the deviations is always zero.

7. The standard deviation is the positive square root of the variance.

The standard deviation and variance can never be negative. Squared deviations can never be negative.

$\{7, 7, 7, 7, 7\} \rightarrow n = 5, \bar{x} = 7$, and $s = 0$

9. When calculating the population standard deviation, you divide the sum of the squared deviations by n, then take the square root of that value. When calculating the sample standard deviation, you divide the sum of the squared deviations by $n - 1$, then take the square root of that value.

When given a data set, one would have to determine if it represented the population or was a sample taken from the population. If the data is a population, then σ is calculated. If the data is a sample, then s is calculated.

11. Range = max − min = 23 − 13 = 10

$\mu = \dfrac{\Sigma x}{N} = \dfrac{232}{14} \approx 16.57$

$\sigma^2 = \dfrac{\Sigma(x - \mu)^2}{N} = \dfrac{143.43}{14} \approx 10.25$

13. Range = max − min = 27 − 8 = 19

$\bar{x} = \dfrac{\Sigma x}{n} = \dfrac{233}{13} \approx 17.92$

$s^2 = \dfrac{\Sigma(x - \bar{x})^2}{(n - 1)} = \dfrac{714.9231}{12} \approx 59.58$

$s = \sqrt{s^2} \approx 7.72$

15. Company B. Due to the larger standard deviation in salaries for company B, it would be more likely to be offered a salary of $33,000.

17. (a) LA: range = max − min = 35.9 − 18.3 = 17.6

$$s^2 = \frac{\Sigma(x - \bar{x})^2}{(n - 1)} = \frac{298.7822}{8} \approx 37.35$$

$$s = \sqrt{s^2} \approx 6.11$$

LB: range = max − min = 26.9 − 18.2 = 8.7

$$s^2 = \frac{\Sigma(x - \bar{x})^2}{(n - 1)} = \frac{298.7822}{8} \approx 8.71$$

$$s = \sqrt{s^2} \approx 2.95$$

(b) It appears from the data that the annual salaries in LA are more variable than the salaries in Long Beach.

19. (a) Pub: range = max − min = 39.9 − 34.8 = 5.1

$$s^2 = \frac{\Sigma(x - \bar{x})^2}{(n - 1)} = \frac{20.675}{7} \approx 2.95$$

$$s = \sqrt{s^2} \approx 1.72$$

Priv: range = max − min = 21.8 − 17.6 = 4.2

$$s^2 = \frac{\Sigma(x - \bar{x})^2}{(n - 1)} = \frac{13.91875}{7} \approx 1.99$$

$$s = \sqrt{s^2} \approx 1.41$$

(b) It appears from the data that the annual salaries for public teachers are more variable than the salaries for private teachers.

21. (a) Greatest sample standard deviation: (ii)
Data set (ii) has more entries that are farther away from the mean.

Least same standard deviation: (iii)
Data set (iii) has more entries that are close to the mean.

(b) The three data sets have the same mean, but have different standard deviations.

23. (a) Similarities: Both estimate proportions of the data contained within k standard deviations of the mean.

Difference: The Empirical Rule assumes the distribution is bell-shaped, Chebychev's Theorem makes no such assumption.

(b) You must know that the distribution is bell-shaped.

(c) If $k = 1$, Chebychev's Theorem would return a proportion equal to zero. If $k < 1$, it would return a negative proportion (which is not possible).

25. $(1000, 1400) \rightarrow (1000, 1000 + 2(200)) \rightarrow (\bar{x}, \bar{x} + 2s)$
47.5% of the farms value between $1000 and $1400 per acre.

27. $1 - \frac{1}{k^2} = \frac{1 - 1}{9} = .8889 \rightarrow$ At least 88.89% of the eruption times lie within 3 standard deviations of the mean.

$(\bar{x} - 3s, \bar{x} + 3s) \rightarrow (.05, 6.59) \rightarrow$ At least 88.89% of the eruption times lie between .05 and 6.59 minutes.

29. $\bar{x} = \Sigma xf/n = 83/40 \approx 2.075$

$$s = \sqrt{\frac{\Sigma(x - \bar{x})^2 f}{n - 1}} = \sqrt{\frac{68.775}{39}} \approx 1.328$$

31.

Class	Midpoint, x	f
0-4	2	19
5-13	9	36
14-17	15.5	15.8
18-24	21	26.3
25-34	29.5	37.2
35-44	39.5	44.7
45-64	54.5	61
65+	70	34.7
		274.7

$$\bar{x} = \frac{\Sigma xf}{n} = \frac{9775.75}{274.7} \approx 35.59$$

$$s = \sqrt{\frac{\Sigma(x - \bar{x})^2 f}{n - 1}} = \sqrt{\frac{123828}{274.7 - 1}} \approx 21.27$$

33. $CV_{\text{heights}} = \frac{\sigma}{\mu} \cdot 100\% = \frac{3.44}{72.75} \cdot 100 \approx 4.73$

$CV_{\text{weights}} = \frac{\sigma}{\mu} \cdot 100\% = \frac{18.47}{187.83} \cdot 100 \approx 9.83$

It appears that weight is more variable than height.

35. (a) $P = \dfrac{3(\bar{x} - \text{median})}{s} = \dfrac{3(17 - 19)}{2.3} \approx -2.61$; skewed left

(b) $P = \dfrac{3(\bar{x} - \text{median})}{s} = \dfrac{3(32 - 25)}{5.1} \approx 4.12$; skewed right

(c) $P = \dfrac{3(\bar{x} - \text{median})}{s} = \dfrac{3(37 - 37)}{6.2} \approx 0$; symmetric

(d) $P = \dfrac{3(\bar{x} - \text{median})}{s} = \dfrac{3(45 - 73)}{10.5} \approx -8$; skewed left

(e) $P = \dfrac{3(\bar{x} - \text{median})}{s} = \dfrac{3(57 - 42)}{9.4} \approx 4.79$; skewed right

37. (a) $\bar{x} = 550$ $\quad s = 302.765$

(b) $\bar{x} = 560$ $\quad s = 302.765$

(c) $\bar{x} = 540$ $\quad s = 302.765$

(d) By adding or subtracting a constant k to each entry, the new sample mean will be $\bar{x} + k$ with the sample standard deviation being unaffected.

2.5   MEASURES OF POSITION

2.5 Try It Yourself Solutions

1a. 0, 1, 1, 1, 2, 2, 2, 3, 3, 4, 4, 4, 5, 5, 5, 6, 6, 6, 6, 7, 7, 8, 8, 9, 10, 10, 10, 11, 11, 11, 12, 12, 13, 15, 16, 16, 17, 17, 21, 21, 22, 23, 24, 25, 25, 26, 27, 27, 28, 28, 29, 30, 31, 31, 32, 32, 33, 33, 34, 36, 39, 41, 42, 45, 46, 47, 48, 49, 50, 50, 51, 52, 53, 54, 55, 56, 63

b. $Q_2 = 21$

c. $Q_1 = 6.5$ $Q_3 = 33.5$

2a. (Enter the data)

b. $Q_1 = 17$ $Q_2 = 23$ $Q_3 = 28.5$

c. One quarter of the tuition costs is below $17,000 or less, one half is $23,000 or less, and three quarters is $28,500 or less.

3a. $Q_1 = 6.5$ $Q_3 = 33.5$

b. $IQR = Q_3 - Q_1 = 33.5 - 6.5 = 27$

c. The ages in the middle half of the data set vary by 27 years.

4a. Min = 0 $Q_1 = 6.5$ $Q_2 = 21$

$Q_3 = 33.5$ Max = 63

bc.

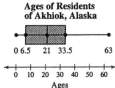

Ages of Residents of Akhiok, Alaska

d. It appears that half of the ages are between 6.5 and 33.5 years.

5a. 85th percentile

b. 85% of the ages are 47 years or younger.

2.5   EXERCISE SOLUTIONS

1. The basketball team scored more points per game than 75% of the teams in the league.

3. The student scored above 63% of the students who took the ACT placement test.

5. True

7. (a) Min = 10

(b) Max = 21

(c) $Q_1 = 13$

(d) $Q_2 = 15$

(e) $Q_3 = 17$

(f) $IQR = 4$

9. (a) Min = 900

   (b) Max = 2100

   (c) $Q_1 = 1250$

   (d) $Q_2 = 1500$

   (e) $Q_3 = 1950$

   (f) $IQR = 700$

11. (a) $Q_1 = 4.5$ $Q_2 = 6$ $Q_3 = 7.5$

   (b)

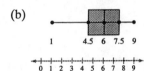

13. (a) $Q_1 = 25$ $Q_2 = 40$ $Q_3 = 47.5$

   (b) $P_{25} = Q_1 = 25$
       $P_{50} = Q_2 = 40$
       $P_{75} = Q_3 = 47.5$

   (c) **Retirements in the House**
       of Representatives

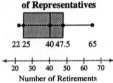

   (d) Half of the retirements are between 25 and 47.5.

15. (a) $Q_1 = 9.85$ $Q_2 = 11.2$ $Q_3 = 13.25$

   (b) $P_{25} = Q_1 = 9.85$
       $P_{50} = Q_2 = 11.2$
       $P_{75} = Q_3 = 13.25$

   (c) **Automotive Mechanics**

   9.4 9.85 11.2   13.25   15.35

   9 10 11 12 13 14 15 16
   Hourly earnings
   (in dollars)

   (d) Half of the hourly earnings are between $9.85 and $13.25.

17. $Q_1 = B,$ $Q_2 = A,$ $Q_3 = C$
   25% of the entries are below B, 50% are below A, and 75% are below C.

19. (a) $Q_1 = 2,$ $Q_2 = 4,$ $Q_3 = 5$

   (b) **Watching Television**

   0   2   4 5   9

   0 1 2 3 4 5 6 7 8 9
   Hours

   (c) Half of the hours of TV watched per day are between 2 and 5.

21. 70 inches
20% of the heights are below 70 inches.

23. (a) $Q_1 = 42, Q_2 = 49, Q_3 = 56$

(b)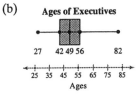

Ages of Executives

(c) Half of the ages are between 42 and 56 years.

(d) About 49 years old ($\bar{x} = 49.62$ and $Q_2 = 49.00$).

CHAPTER 2 REVIEW EXERCISE SOLUTIONS

1.

Class	Midpoint	Boundaries	Frequency	Relative Frequency	Cumulative Frequency
20–23	21.5	19.5–23.5	1	0.05	1
24–27	25.5	23.5–27.5	2	0.10	3
28–31	29.5	27.5–31.5	6	0.30	9
32–35	33.5	31.5–35.5	7	0.35	16
36–39	37.5	35.5–39.5	4	0.20	20
			20	1	

3. (See problem 1)

5.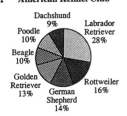

Liquid Volume 12 oz Cans

7.

Meals Purchased

9. Average Daily Highs

11. American Kennel Club

13.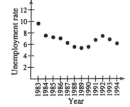

U.S. Unemployment Rate

15. $\bar{x} = 30.8$ median $= 30$ mode $= 29$

17. $\bar{x} = \dfrac{\Sigma xf}{n} = \dfrac{125}{60} = 2.083$

19. $\bar{x} = \dfrac{\Sigma xw}{w} = \dfrac{8740}{100} = 87.4$

21. Skewed

23. Skewed right

25. Mean

27. Range $= $ Max $-$ Min $= 19.73 - 15.89 = 3.84$

29. $\mu = \dfrac{\Sigma x}{N} = 63.67$

$\sigma = \sqrt{\dfrac{\Sigma(x - \mu)^2}{N}} = 8.11$

31. $\bar{x} = \dfrac{\Sigma x}{n} = 38{,}653.5$

$s = \sqrt{\dfrac{\Sigma(x - \bar{x})^2}{n - 1}} = 6762.2$

33. $(\$24.50, \$30.00) \to (\mu, \mu + 2\sigma) \to$ The percent of the rates between \$24.50 and \$30.00 is 47.5%.

35. $1 - \dfrac{1}{k^2} = .89 \to$ At least 89% of the flight lengths is contained within 3 standard deviations of the mean
$(\mu - 3\sigma, \mu + 3\sigma) \to (6.35 - 3 \cdot 2.08, 6.35 + 3 \cdot 2.08) = (0.11, 12.59)$
The percent of the flight lengths between 0.11 and 12.59 days is at least 89%.

37. $\bar{x} = \dfrac{\Sigma xf}{n} = \dfrac{61}{25} = 2.44$

$s = \sqrt{\dfrac{\Sigma(x - \bar{x})^2 f}{n - 1}} = \sqrt{\dfrac{72.16}{24}} \approx 1.73$

39. $Q_3 = 70$

41.

Height of Students

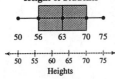

43.

Weight of Football Players

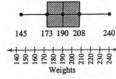

45. $84/753 \approx .11155 \to 11\%$ have larger audiences. The station would represent the 89th percentile, P_{89}.

CHAPTER 2 QUIZ SOLUTIONS

1. (a)

Class Limits	Midpoint	Class Boundaries	Frequency	Relative Frequency	Cumulative Frequency
101–112	106.5	100.5–112.5	3	0.12	3
113–124	118.5	112.5–124.5	11	0.44	14
125–136	130.5	124.5–136.5	7	0.28	21
137–148	142.5	136.5–148.5	2	0.08	23
149–160	154.5	148.5–160.5	2	0.08	25

(b) Frequency Histogram and Polygon

(c) Relative Frequency Histogram

(d) Skewed

(e)

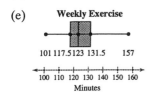

(f)

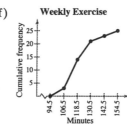

2. $\bar{x} = \dfrac{\Sigma xf}{n} = \dfrac{3130.5}{25} \approx 125.22$

$s = \sqrt{\dfrac{\Sigma(x - \bar{x})^2 f}{n - 1}} = \sqrt{\dfrac{4055.04}{24}} \approx 13.00$

3. (a)

(b)

4. (a) $\bar{x} = \dfrac{\Sigma x}{n} = 751.63$

median = 784.5

mode = (none)

(b) range = Max − min = 575

$s^2 = \dfrac{\Sigma(x - \bar{x})^2}{n - 1} = 48{,}135.13$

$s = \sqrt{\dfrac{\Sigma(x - \bar{x})^2}{n - 1}} = 219.40$

5. $\bar{x} - 2s = 155{,}000 - 2 \cdot 15{,}000 = 125{,}000$
    $\bar{x} + 2s = 155{,}000 + 2 \cdot 15{,}000 = 185{,}000$

6. (a) $Q_1 = 76$ $Q_2 = 79$ $Q_3 = 88$

   (b) $IRQ = Q_3 - Q_1 = 12$

   (c)   **Wins for Each Team**

Probability

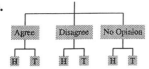

3.1 BASIC CONCEPTS OF PROBABILITY

3.1 Try It Yourself Solutions

1ab.

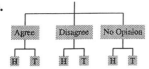

c. 6 outcomes

d. Let A = Agree, D = Disagree, N = No Opinion, H = Heads and T = Tails
Sample space = $\{AH, AT, DH, DT, NH, NT\}$

2a. (1) 6 outcomes   (2) 1 outcome

b. (1) Not a simple event   (2) Simple event

3a. (1) 52   (2) 1   (3) $P(7 \text{ of diamonds}) = \dfrac{1}{52} \approx 0.0192$

b. (1) 52   (2) 13   (3) $P(\text{diamond}) = \dfrac{13}{52} = 0.25$

c. (1) 52   (2) 52   (3) $P(\text{diamond, heart, club, or spade}) = \dfrac{52}{52} = 1$

4a. Event = a fraudulent claim is found (Freq = 4)

b. Total Frequency = 100

c. $P(\text{fraudulent claim}) = \dfrac{4}{100} = 0.04$

5a. Frequency = 54

b. Total of the Frequencies = 1000

c. $P(\text{age 15 to 24}) = \dfrac{54}{1000} = 0.054$

6a. Event = salmon successfully passing through a dam on the Columbia River.

b. Experimentation

c. Empirical probability

7a. $P(\text{red gill}) = \dfrac{17}{40} = 0.425$

b. $P(\text{not red gill}) = 1 - \dfrac{17}{40} = \dfrac{23}{40} = 0.575$

c. $\dfrac{23}{40}$ or 0.575

3.1 EXERCISE SOLUTIONS

1. (a) Yes, the probability of an event occurring must be contained in the interval [0, 1] or [0%, 100%].

(b) No, the probability of an event occurring cannot be larger than 1.

(c) No, the probability of an event occurring cannot be smaller than 0.

(d) Yes, the probability of an event occurring must be contained in the interval $[0, 1]$ or $[0\%, 100\%]$

(e) Yes, the probability of an event occurring must be contained in the interval $[0, 1]$ or $[0\%, 100\%]$

3. $\{0, 1, 2, 3, 4, 5, 6, 7, 8, 9\}$

5. $\{(A, M), (A, F), (B, M), (B, F),(AB, M), (AB, F), (O, M), (O, F)\}$where (A, M) represents a male with blood type A, (A, F) represents a female with blood type A, etc.

7. Simple event because it is an event that consists of a single outcome.

9. Empirical probability since company records were probably used to calculate the frequency of a washing machine breaking down.

11. $P(\text{did not vote for Bill Clinton}) = \dfrac{48{,}872{,}000}{96{,}273{,}000} \approx 0.508$

13. $P(\text{less than 1000}) = \dfrac{999}{6296} \approx 0.159$

15. $P(\text{number divisible by 1000}) = \dfrac{6}{6296} \approx 0.000953$

17. $P(\text{between 21 and 24}) = \dfrac{13.9}{193.7} \approx 0.072$

19. $P(\text{not between 18 and 20}) = \dfrac{1 - 10.8}{193.7} \approx 0.793$

21. (a) $P(\text{pink}) = \dfrac{2}{4} = 0.5$ (b) $P(\text{red}) = \dfrac{1}{4} = 0.25$ (c) $P(\text{white}) = \dfrac{1}{4} = 0.25$

23. $P(\text{service industry}) = \dfrac{95{,}520}{127{,}900} \approx 0.747$

25. $P(\text{not in service industry}) = 1 - P(\text{service industry}) = 1 - 0.747 = 0.253$

27. The probability of choosing a tea drinker who does not have a college degree.

29. (a) $P(\text{at least 21}) = \dfrac{39}{77} \approx 0.506$ (b) $P(\text{between 40 and 50 inclusive}) = \dfrac{9}{77} \approx 0.117$

   (c) $P(\text{older than 65}) = \dfrac{0}{77} = 0$

31. No, the odds of winning a prize are 1:5. (One winning cap and 5 losing caps)

33. $13:39 = 1:3$

3.2 CONDITIONAL PROBABILITY AND THE MULTIPLICATION RULE

3.2 Try It Yourself Solutions

1a. (1) 30 and 102   (2) 11 and 50

b. (1) $P(\text{not have gene}) = \dfrac{30}{102} \approx 0.294$ (2) $P(\text{not have gene}|\text{normal IQ}) = \dfrac{11}{50} = 0.22$

2a. (1) No   (2) Yes

b. (1) Independent   (2) Dependent

c. (1) A salmon successfully swimming through 1st dam does not affect the probability of successfully swimming through the 2nd dam.

(2) It has been shown in studies that exercising frequently lowers the resting rate of the heart.

3a. (1) Independent   (2) Dependent

b. (1) Let A = {swimming through 1st dam}
B = {swimming through 2nd dam}
$$P(A \text{ and } B) = P(A) \cdot P\left(\dfrac{B}{A}\right) = (0.85) \cdot (0.85) = 0.7225$$

(2) Let A = {does not have gene}
B = {normal IQ}
$$P(A \text{ and } B) = P(A) \cdot P\left(\dfrac{B}{A}\right) = \left(\dfrac{30}{102}\right) \cdot \left(\dfrac{11}{30}\right) \approx 0.108$$

4a. (1) Find probability of the event   (2) Find probability of the compliment of the event

b. (1) $P(3 \text{ salmon successful}) = (0.90) \cdot (0.90) \cdot (0.90) = 0.729$

(2) $P(\text{at least one salmon successful}) = 1 - P(\text{non are successful})$
$$= 1 - (0.10) \cdot (0.10)(0.10) = 0.999$$

3.2 EXERCISE SOLUTIONS

1. Two events are independent if the occurrence of one of the events does not affect the probability of the occurrence of the other event.

If $P\left(\dfrac{B}{A}\right) = P(B)$ or $P\left(\dfrac{A}{B}\right) = P(A)$, then Events A and B are independent.

3. False. If two events are independent, $P\left(\dfrac{A}{B}\right) = P(A)$.

5. These events are independent since the outcome of the 1st card drawn does not affect the outcome of the 2nd card drawn.

7. These events are dependent since the outcome of the 1st ball drawn affects the outcome of the 2nd ball drawn.

9. Let A = {have mutated BRCA gene} and B = {develop breast cancer}. Thus
$$P(B) = \dfrac{1}{9}, P(A) = \dfrac{1}{250}, \text{ and } P\left(\dfrac{B}{A}\right) = \dfrac{8}{10}.$$

(a) $P\left(\dfrac{B}{A}\right) = \dfrac{8}{10} = 0.8$

(b) $P(A \text{ and } B) = P(A) \cdot P\left(\dfrac{B}{A}\right) = \left(\dfrac{1}{250}\right) \cdot \left(\dfrac{8}{10}\right) = 0.0032$

(c) Dependent since $P\left(\dfrac{B}{A}\right) \neq P(B)$.

11. Let $A = \{\text{pregnant}\}$ and $B = \{\text{multiple births}\}$. Thus $P(A) = 0.24$ and $P\left(\dfrac{B}{A}\right) = 0.07$.

(a) $P(A \text{ and } B) = P(A) \cdot P\left(\dfrac{B}{A}\right) = (0.24) \cdot (0.07) = 0.0168$

(b) $P\left(\dfrac{B'}{A}\right) = 1 - P\left(\dfrac{B}{A}\right) = 1 - 0.07 = 0.93$

13. Let $A = \{\text{1st part drawn is defective}\}$ and $B = \{\text{2nd part drawn is defective}\}$

(a) $P(A \text{ and } B) = P(A) \cdot P\left(\dfrac{B}{A}\right) = \left(\dfrac{4}{11}\right) \cdot \left(\dfrac{3}{10}\right) \approx 0.109$

(b) $P(A' \text{ and } B') = P(A') \cdot P\left(\dfrac{B'}{A'}\right) = \left(\dfrac{7}{11}\right) \cdot \left(\dfrac{6}{10}\right) \approx 0.382$

(c) $P(\text{at least one part defective}) = 1 - P(\text{both are not defective}) = 1 - 0.382 = 0.618$.

15. Let $A = \{\text{have one month's income or more}\}$ and $B = \{\text{man}\}$

(a) $P(A) = \dfrac{172}{205} \approx 0.839$

(b) $P\left(\dfrac{A'}{B}\right) = \dfrac{17}{102} \approx 0.167$

(c) $P\left(\dfrac{B'}{A}\right) = \dfrac{87}{172} \approx 0.506$

(d) Dependent since $P(A') \approx 0.161 \neq 0.167 \approx P(A'|B)$

17. (a) $P(\text{all five have AB}+) = (0.03) \cdot (0.03) \cdot (0.03) \cdot (0.03) \cdot (0.03) = 0.0000000243$

(b) $P(\text{none have AB}+) = (0.97) \cdot (0.97) \cdot (0.97) \cdot (0.97) \cdot (0.97) \approx 0.859$

(c) $P(\text{at least one has AB}+) = 1 - P(\text{none have AB}+) = 1 - 0.859 = 0.141$

19. (a) $P(\text{first question correct}) = 0.2$

(b) $P(\text{first two questions correct}) = (0.2) \cdot (0.2) = 0.04$

(c) $P(\text{first three questions correct}) = (0.2)^3 = 0.008$

(d) $P(\text{none correct}) = (0.8)^3 = 0.512$

(e) $P(\text{at least one correct}) = 1 - P(\text{none correct}) = 1 - 0.512 = 0.488$

21. Let $A = \{\text{flight departs on time}\}$ and $B = \{\text{flight arrives on time}\}$

$P\left(\dfrac{A}{B}\right) = \dfrac{P(A \text{ and } B)}{P(B)} = \dfrac{(0.83)}{(0.87)} \approx 0.954$

23. (a) $P\left(\dfrac{A}{B}\right) = \dfrac{P(A) \cdot P\left(\dfrac{B}{A}\right)}{P(A) \cdot P\left(\dfrac{B}{A}\right) + P(A') \cdot P\left(\dfrac{B}{A'}\right)} = \dfrac{\left(\dfrac{2}{3}\right) \cdot \left(\dfrac{1}{5}\right)}{\left(\dfrac{2}{3}\right) \cdot \left(\dfrac{1}{5}\right) + \left(\dfrac{1}{3}\right) \cdot \left(\dfrac{1}{2}\right)} = 0.444$

(b) $P\left(\dfrac{A}{B}\right) = \dfrac{P(A) \cdot P\left(\dfrac{B}{A}\right)}{P(A) \cdot P\left(\dfrac{B}{A}\right) + P(A') \cdot P\left(\dfrac{B}{A'}\right)} = \dfrac{\left(\dfrac{3}{8}\right) \cdot \left(\dfrac{2}{3}\right)}{\left(\dfrac{3}{8}\right) \cdot \left(\dfrac{2}{3}\right) + \left(\dfrac{5}{8}\right) \cdot \left(\dfrac{3}{5}\right)} = 0.4$

25. (a) $P(\text{different birthdays}) = \dfrac{364}{365} \cdot \dfrac{363}{365} \cdot \cdots \cdot \dfrac{342}{365} \approx 0.462$

(b) $P(\text{at least two have same birthday}) = 1 - P(\text{different birthdays}) = 1 - 0.462 = 0.538$

(c) Yes, there were 2 birthdays on the 118th day.

(d) Answers will vary.

3.3 THE ADDITION RULE

3.3 Try It Yourself Solutions

1a. (1) None of the statements are true.

(2) None of the statements are true.

(3) Events A and B cannot occur at the same time.

b. (1) A and B are not mutually exclusive.

(2) A and B are not mutually exclusive.

(3) A and B are mutually exclusive.

2a. (1) Mutually exclusive   (2) Not mutually exclusive

b. (1) Let $A = \{6\}$ and $B = \{\text{odd}\}$

$P(A) = \dfrac{1}{6}$ and $P(B) = \dfrac{3}{6} = \dfrac{1}{2}$

(2) Let $A = \{\text{face card}\}$ and $B = \{\text{heart}\}$

$P(A) = \dfrac{12}{52}$, $P(B) = \dfrac{13}{52}$, and $P(A \text{ and } B) = \dfrac{3}{52}$

c. (1) $P(A \text{ or } B) = P(A) + P(B) = \dfrac{1}{6} + \dfrac{1}{2} \approx 0.667$

(2) $P(A \text{ or } B) = P(A) + P(B) - P(A \text{ and } B) = \dfrac{12}{52} + \dfrac{13}{52} - \dfrac{3}{52} \approx 0.423$

3a. $A = \{\text{sales between \$0 and \$24,999}\}$

$B = \{\text{sales between \$25,000 and \$49,000}\}$

b. A and B cannot occur at the same time $\rightarrow A$ and B are mutually exclusive

c. $P(A) = \dfrac{3}{36}$ and $P(B) = \dfrac{5}{36}$

d. $P(A \text{ or } B) = P(A) + P(B) = \dfrac{3}{36} + \dfrac{5}{36} \approx 0.222$

4a. (1) Mutually exclusive   (2) Not mutually exclusive

b. (1) $P(B \text{ or } AB) = P(B) + P(AB) = \dfrac{45}{409} + \dfrac{16}{409} \approx 0.149$

(2) $P(O \text{ or } Rh+) = P(O) + P(Rh+) - P(O \text{ and } Rh+) = \dfrac{184}{409} + \dfrac{344}{409} - \dfrac{156}{409} \approx 0.910$

5a. Let $A = \{\text{linebacker}\}$ and $B = \{\text{quarterback}\}$

$P(A \text{ or } B) = P(A) + P(B) = \dfrac{34}{241} + \dfrac{8}{241} \approx 0.174$

b. $P(\text{not a linebacker or quarterback}) = 1 - P(A \text{ or } B) = 1 - 0.174 = 0.826$

3.3 EXERCISE SOLUTIONS

1. $P(A \text{ and } B) = 0$ because A and B cannot occur at the same time.

3. True

5. Not mutually exclusive since the two events can occur at the same time.

7. Not mutually exclusive since the two events can occur at the same time.

9. Mutually exclusive since the two events cannot occur at the same time.

11. (a) No, it is possible for the events {overtime} and {temporary help} to occur at the same time.

(b) $P(\text{OT or temp}) = P(\text{OT}) + P(\text{temp}) - P(\text{OT and temp}) = \dfrac{18}{52} + \dfrac{9}{52} - \dfrac{5}{52} \approx 0.423$

13. (a) Not mutually exclusive since the two events can occur at the same time.

(b) $P(\text{puncture or corner}) = P(\text{puncture}) + P(\text{corner}) - P(\text{puncture and corner})$
$= 0.05 + 0.08 - 0.004 = 0.126$

15. (a) $P(\text{under 5}) = 0.069$

(b) $P(\text{not 65+}) = 1 - P(65+) = 1 - 0.126 = 0.874$

(c) $P(\text{between 18 and 34}) = P(18 - 24 \text{ or } 25 - 34) = P(18 - 24) + P(25 - 34)$
$= 0.096 + 0.136 = 0.232$

17. (a) $P(\text{1st LH and 2nd LH}) = \left(\dfrac{120}{1000}\right) \cdot \left(\dfrac{119}{999}\right) \approx 0.014$

(b) $P(\text{at least one is LH}) = 1 - P(\text{neither are LH}) = 1 - \left(\dfrac{880}{1000}\right) \cdot \left(\dfrac{879}{999}\right) \approx 0.226$

(c) $P(\text{neither are LH}) = \left(\dfrac{880}{1000}\right) \cdot \left(\dfrac{879}{999}\right) \approx 0.774$

(d) (b) and (c) are complimentary events since {neither are LH} is the set of all outcomes in the sample space that are not included in {at least one is LH}.

19. Answers will vary.

Conclusion: If two events, $\{A\}$ and $\{B\}$, are independent, $P(A \text{ and } B) = P(A) \cdot P(B)$. If two events are mutually exclusive, $P(A \text{ and } B) = 0$. The only scenario when two events can be independent and mutually exclusive is if $P(A) = 0$ or $P(B) = 0$.

3.4 COUNTING PRINCIPLES

3.4 Try It Yourself Solutions

1a. Manufacturer: 4
Size: 3
Color: 6

b. $4 \cdot 3 \cdot 6 = 72$ ways

2a. (1) Each letter is an event (26 choices)
(2) Each letter is an event (26, 25, 24, 23, 22, and 21 choices)

b. (1) $26 \cdot 26 \cdot 26 \cdot 26 \cdot 26 \cdot 26 = 26^6 = 308{,}915{,}776$

(2) $26 \cdot 25 \cdot 24 \cdot 23 \cdot 22 \cdot 21 = 165{,}765{,}600$

3a. $n = 6$ teams   **b.** $6! = 720$

4a. $_8P_3 = \dfrac{8!}{(8-3)!} = \dfrac{8!}{5!} = \dfrac{8 \cdot 7 \cdot 6 \cdot 5 \cdot 4 \cdot 3 \cdot 2 \cdot 1}{5 \cdot 4 \cdot 3 \cdot 2 \cdot 1} = 8 \cdot 7 \cdot 6 = 336$

b. There are 336 possible ways that three horses can finish in first, second, and third place.

5a. $n = 12, r = 4$ **b.** $_{12}P_4 = \dfrac{12!}{(12-4)!} = \dfrac{12!}{8!} = 11{,}880$

6a. $n = 20, n_1 = 6, n_2 = 9, n_3 = 5$ **b.** $\dfrac{n!}{n_1! \cdot n_2! \cdot n_3!} = \dfrac{20!}{6! \cdot 9! \cdot 5!} = 77{,}597{,}520$

7a. $n = 16, r = 3$

b. $_{16}C_3 = 560$

c. There are 560 different possible 3 person committees that can be selected from 16 employees.

8a. 1 favorable outcome and $\dfrac{6!}{1! \cdot 2! \cdot 2! \cdot 1!} = 180$ distinguishable permutations.

b. $P(\text{Letter}) = \dfrac{1}{180} \approx 0.0056$

9a. $(_5C_3) \cdot (_7C_{10}) = 10 \cdot 1 = 10$ **b.** $_{12}C_3 = 220$ **c.** $\dfrac{10}{220} \approx 0.045$

3.4 EXERCISE SOLUTIONS

1. You are counting the number of ways two or more events can occur in sequence.

3. False, a permutation is an ordered arrangement of objects.

5. Permutation, since order of the ten people in line matters.

7. $10 \cdot 8 \cdot _{13}C_2 = 6240$

9. $9 \cdot 10 \cdot 10 \cdot 5 = 4500$

11. $8! = 40{,}320$

13. $10! = 3{,}628{,}800$

15. $_{10}P_3 = 720$

17. $_{15}P_4 = 32,760$

19. $\dfrac{18!}{4! \cdot 8! \cdot 6!} = 9,189,180$

21. $_{40}C_{12} = 5,586,853,480$

23. $_8C_3 = 56$

25. (a) $_8C_4 = 70$ (b) $2 \cdot 2 \cdot 2 \cdot 2 = 16$

27. (a) $(_8C_3) \cdot (_2C_0) = (56) \cdot (1) = 56$

  (b) $(_8C_2) \cdot (_2C_1) = (28) \cdot (2) = 56$

  (c) At least two good units = 1 defective of 0 defective units
    $56 + 56 = 112$

29. (a) $_{40}C_5 = 658,008$ (b) $P(\text{win}) = \dfrac{1}{658,008} \approx 0.00000152$

31. $P(\text{all 4 have no interest}) = \dfrac{(_{21}C_4)}{(_{100}C_4)} \approx 1.067 \times 10^{-8}$

CHAPTER 3 REVIEW EXERCISE SOLUTIONS

1. Sample space:

  {HHHH, HHHT, HHTH, HHTT, HTHH, HTHT, HTTH, HTTT, THHH, THHT, THTH, THTT, TTHH, TTHT, TTTH, TTTT}

  Event: Getting three heads
    {HHHT, HHTH, HTHH, THHH}

3. Empirical probability

5. Subjective probability

7. Classical probability

9. $P(\text{at least 20}) = 1 - P(0 - 19) = 1 - 0.29 = 0.71$

11. $P(\text{undergrad} \mid +) = 0.92$

13. $P(\text{home or work}) = P(\text{home}) + P(\text{work}) - P(\text{home and work}) = 0.44 + 0.37 - 0.21 = 0.60$

15. Independent, the first event does not affect the outcome of the second event.

17. $P(\text{correct toothpaste and correct dental rinse}) = P(\text{correct toothpaste}) \cdot P(\text{correct dental rinse}) = \left(\dfrac{1}{6}\right) \cdot \left(\dfrac{1}{4}\right) \approx 0.0417$

19. Mutually exclusive since both events cannot occur at the same time.

21. $\dfrac{20 + 13 - 5}{52} \approx 0.538$

23. $8 \cdot 2 \cdot 9 = 144$

25. $\dfrac{9!}{6! \cdot 3!} = 84$

27. Order is important: $_{15}P_3 = 2730$

29. Order is not important: $_{17}C_4 = 2380$

31. (a) $P(\text{no defectives}) = \dfrac{_{197}C_3}{_{200}C_3} = \dfrac{1{,}254{,}890}{1{,}313{,}400} \approx 0.955$

   (b) $P(\text{all defective}) = \dfrac{_3C_3}{_{200}C_3} = \dfrac{1}{1{,}313{,}400} = \; \approx 0.000000761$

   (c) $P(\text{at least one defective}) = 1 - P(\text{no defective}) = 1 - 0.955 = 0.045$

   (d) $P(\text{at least one non-defective}) = 1 - P(\text{all defective}) = 1 - 0.000000761 - 0.999999239$

CHAPTER 3 QUIZ SOLUTIONS

1. (a) $P(\text{bachelor}) = \dfrac{1161}{2153} \approx 0.539$ (b) $P(\text{bachelor}|F) = \dfrac{659}{1228} \approx 0.537$

   (c) $P(\text{bachelor}|M) = \dfrac{502}{925} \approx 0.543$ (d) $P(\text{associate or bachelor}) = \dfrac{532 + 1161}{2153} \approx 0.786$

   (e) $P(\text{doctorate}|M) = \dfrac{27}{925} \approx 0.0292$

2. Not mutually exclusive since both events can occur at the same time.
Dependent since one event can affect the occurrence of the second event.

3. (a) $_{147}C_3 = 518{,}665$ (b) $(_{147}C_2) \cdot (_3C_1) = 32{,}193$ (c) $518{,}665 + 32{,}193 = 550{,}858$

4. $9 \cdot 10 \cdot 10 \cdot 5 = 4500$

5. $_{25}P_4 = 303{,}600$

CUMULATIVE TEST SOLUTIONS FOR CHAPTERS 1–3

1. Quantitative, Ratio

2. Use the sampling method of data collection. The sampling technique should be a simple random sample because it would be difficult to collect this information from the entire population of students.

3.

Class Limits	Midpoint	Freq	Boundaries	Rel Freq	Cum Freq
90-110	100	9	89.5-110.5	0.300	9
111-131	121	8	110.-131.5	0.267	17
132-152	142	3	131.5-152.5	0.100	20
153-173	163	5	152.5-173.5	0.167	25
174-195	184	5	173.5-195.5	0.167	30

4.

Book Expenses

5.

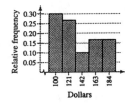

Book Expenses

6. Key: $10|3 = 103$

```
 9 | 0 1 3 8
10 | 3 4 9
11 | 0 0 1 6 7 8 9
12 | 0 3 7
13 | 2 6
14 |
15 | 0 3 6
16 | 0 2
17 | 0 8
18 | 1 7
19 | 1 5
```

7. Skewed

8.

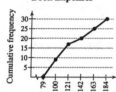

Book Expenses

9. $\bar{x} = \dfrac{\Sigma x}{n} = \dfrac{4010}{30} \approx 133.7$

median = 121.5
mode = 110
These values are statistics since they are characteristics of a sample.

10. range = max − min = 195 − 90 = 105

$$s^2 = \frac{\Sigma(x - \bar{x})^2}{n - 1} \approx 1040.506$$

$$s = \sqrt{\frac{\Sigma(x - \bar{x})^2}{n - 1}} \approx 32.257$$

The sample standard deviation is $32.26.

11. $Q_1 = 110 \qquad Q_2 = 121.5 \qquad Q_3 = 160$
$IQR = Q_3 - Q_1 = 150$

Book Expenses

90 110 121.5 160   195

```
80  100  120  140  160  180  200
```
Dollars

12. $P(\text{less than }\$120) = \dfrac{14}{30} = 0.467 \qquad P(\text{more than }\$120) = \dfrac{15}{30} = 0.5$

13. $P(\text{less than }\$120 \text{ or more than }\$160) = P(\text{less than }\$120) + P(\text{more than }\$160) = \dfrac{14}{30} + \dfrac{7}{30} = 0.7$

14. P(at least one spent more than $175) = 1 − P(none spent more than $175

$$= 1 - \left(\frac{25}{30} \cdot \frac{24}{29} \cdot \frac{23}{28} \cdot \frac{22}{27} \cdot \frac{21}{26}\right) \approx 0.627$$

15. Order is not important: $_{30}C_5 = 142{,}506$

Discrete Probability Distributions

4.1 Try It Yourself Solutions

Section 4.1

1a. (1) measured   (2) counted

  b. (1) Random variable is continuous since x can be any amount of time needed to complete a test.

    (2) Random variable is discrete since x can be counted.

2ab.

x	f	$P(x)$
0	16	0.16
1	19	0.19
2	15	0.15
3	21	0.21
4	9	0.09
5	10	0.10
6	8	0.08
7	2	0.02
	100	1

c.

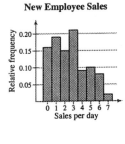

New Employee Sales

3a. See solution for 2a. Each $P(x)$ is between 0 and 1.

  b. See solution for 2b. $\Sigma P(x) = 1$

  c. (1) Since both conditions were not met, the distribution is not a probability distribution.
    (2) Since both conditions are met, the distribution is a probability distibution.

4a. (1) Yes, each outcome is between 0 and 1.   (2) Yes, each outcome is between 0 and 1.

  b. (1) No, $\Sigma P(x) = \dfrac{18}{16} \neq 1$ (2) Yes, $\Sigma P(x) = 1$

  c. (1) Not a probability distribution   (2) Is a probability distribution

5ab.

x	$P(x)$	$xP(x)$
0	0.16	0.00
1	0.19	0.19
2	0.15	0.30
3	0.21	0.63
4	0.09	0.36
5	0.10	0.50
6	0.08	0.48
7	0.02	0.14
	1	2.60

  c. $\mu = \Sigma xP(x) = 2.60$

    On average, 2.60 sales are made per day.

6ab.

x	$P(x)$	$x - \mu$	$(x - \mu)^2$	$P(x)(x - \mu)^2$
0	0.16	−2.6	6.76	1.0816
1	0.19	−1.6	2.56	0.4864
2	0.15	−0.6	0.36	0.054
3	0.21	0.4	0.16	0.0336
4	0.09	1.4	1.96	0.1764
5	0.10	2.4	5.76	0.576
6	0.08	3.4	11.56	0.9248
7	0.02	4.4	19.36	0.3872
	$\Sigma P(x) = 1$			$\Sigma P(x)(x - \mu)^2 = 3.72$

c. $\sigma = \sqrt{\sigma^2} = \sqrt{3.720} \approx 1.93$

7ab.

x	f	$P(x)$	$xP(x)$
0	25	0.11	0.000
1	48	0.213	0.213
2	60	0.267	0.533
3	45	0.200	0.600
4	20	0.089	0.356
5	10	0.044	0.222
6	8	0.036	0.213
7	5	0.022	0.156
8	3	0.013	0.107
9	1	0.004	0.040
	225	1	2.440

c. $E(x) = \Sigma xP(x) = 2.44$

d. You can expect an average of 2.44 sales per day.

4.1 EXERCISE SOLUTIONS

1. A random variable represents a numerical value assigned to an outcome of a probability experiment.

Examples: Answers will vary.

3. False. In most applications, discrete random variables represent counted data, while continuous random variables represent measured data.

5. True

7. Discrete because home attendance is a random variable that is countable.

9. Discrete because the random variable is countable.

11. Continuous because the random variable has an infinite number of possible outcomes and cannot be counted.

13. Discrete because the random variable is countable.

15. $\Sigma P(x) \rightarrow P(3) = 0.22$

17. Yes

19. No, $\Sigma P(x) = 0.95$ and $P(5) < 0$.

21. (a) $\mu = \Sigma xP(x) \approx 2.1$ (b) $\sigma^2 = \Sigma(x - \mu)^2 P(x) \approx 1.09$ (c) $\sigma = \sqrt{\sigma^2} = 1.044$

23. (a)

x	f	P(x)	xP(x)	(x − μ)²P(x)
0	316	0.316	0	0.3761
1	425	0.425	0.425	0.0035
2	168	0.168	0.336	0.1388
3	48	0.048	0.144	0.1749
4	29	0.029	0.116	0.2454
5	14	0.014	0.07	0.2139
	1000	1	1.091	1.1527

(b) $\mu = \Sigma xP(x) \approx 1.091$ (c) $\sigma^2 = \Sigma(x - \mu)^2P(x) \approx 1.153$ (d) $\sigma = \sqrt{\sigma^2} \approx 1.074$

(e) A household on average has 1.091 dogs with a standard deviation of 1.074.

25. (a)

x	f	P(x)	xP(x)	(x − μ)²P(x)
0	300	0.432	0.000	0.252
1	280	0.403	0.403	0.022
2	95	0.137	0.274	0.209
3	20	0.029	0.087	0.145
	695	1	0.764	0.629

(b) $\mu = \Sigma xP(x) \approx 0.764$ (c) $\sigma^2 = \Sigma(x - \mu)^2P(x) \approx 0.629$ (d) $\sigma = \sqrt{\sigma^2} \approx 0.793$

27. (a) $\mu = \Sigma xP(x) \approx 18.375$ (b) $\sigma^2 = \Sigma(x - \mu)^2P(x) \approx 41.734$

(c) $\sigma = \sqrt{\sigma^2} \approx 6.460$ (d) $E(x) = \mu = \Sigma xP(x) \approx 18.375$

(e) The publisher can anticipate an average of $72,581.25 (18,375 × $3.95) per week to be generated by magazine sales.

29. (a) $P(x > 2) = 0.25 + 0.10 = 0.35$ (b) $P(x < 4) = 1 - P(4) = 1 - 0.10 = 0.90$

31. (a) $P(x < 2) = 0.316 + 0.425 = 0.741$ (b) $P(x \geq 2) = 1 - P(x < 2) = 1 - 0.741 - 0.259$

(c) $P(2 \leq x \leq 4) = 0.168 + 0.048 + 0.029 = 0.245$

33. $E(x) = \mu = \Sigma xP(x) = (-1) \cdot \left(\frac{37}{38}\right) + (35) \cdot \left(\frac{1}{38}\right) \approx \$0.05$

4.2 BINOMIAL DISTRIBUTIONS

4.2 Try It Yourself Solutions

1a. Trial: Individual questions (10 trials)
Success: question answered correctly

b. Yes, the experiment satisfies the four conditions of a binomial experiment.

c. $n = 10, p = 0.25, q = 0.75, x = 0, 1, 2, \ldots, 9, 10$

2a. Trial: 5 cards being drawn with replacement
Success: card drawn is a club
Failure: card drawn is not a club

b. $n = 5, p = 0.25, q = 0.75, x = 3$

c. $P(3) = \frac{5!}{2!3!}(0.25)^3(0.75)^2 \approx 0.088$

3a. Trial: 7 retirees
Success: Selecting a retiree who responded "yes"
Failure: Selecting a retiree who responded "no"

b. $n = 7, p = 0.71, q = 0.29, x = 0, 1, 2, \ldots, 6, 7$

c. $P(0) = {}_7C_0(0.71)^0(0.29)^7 = 0.000172$
$P(1) = {}_7C_1(0.71)^1(0.29)^6 = 0.00296$
$P(2) = {}_7C_2(0.71)^2(0.29)^5 = 0.0217$
$P(3) = {}_7C_3(0.71)^3(0.29)^4 = 0.0886$
$P(4) = {}_7C_4(0.71)^4(0.29)^3 = 0.217$
$P(5) = {}_7C_5(0.71)^5(0.29)^2 = 0.319$
$P(6) = {}_7C_6(0.71)^6(0.29)^1 = 0.260$
$P(7) = {}_7C_7(0.71)^7(0.29)^0 = 0.091$

d.

x	$P(x)$
0	0.000172
1	0.00296
2	0.0217
3	0.0886
4	0.217
5	0.319
6	0.260
7	0.091

4a. Trial: 10 businesses
Success: Selecting a business with a Web site
Failure: Selecting a business with out a site

b. $n = 10, p = 0.25, x = 4$ **c.** $P(4) \approx 0.146$

5a. $n = 250, p = 0.71, x = 178$ **b.** $P(178) \approx 0.056$

c. The probability that exactly 178 Americans will use more than one topping on their hotdog is about 0.056.

6a. (1) $x = 2$ (2) $x = 2, 3, 4,$ or 5   (3) $x = 0$ or 1

b. (1) $P(2) \approx 0.217$

(2) $P(x \geq 2) = 1 - P(0) - P(1) = 1 - 0.308 - 0.409 = 0.283$

(3) $P(x < 2) = P(0) + P(1) = 0.308 + 0.409 = 0.717$

c. (1) The probability that exactly two men consider fishing their favorite leisure-time activity is about 0.217.

(2) The probability that at least two men consider fishing their favorite leisure-time activity is about 0.283.

(3) The probability that fewer than two men consider fishing their favorite leisure-time activity is about 0.717.

7a. $P(0) = {}_6C_0(0.41)^0(0.59)^6 = 0.042$
$P(1) = {}_6C_1(0.41)^1(0.59)^5 = 0.176$
$P(2) = {}_6C_2(0.41)^2(0.59)^4 = 0.306$
$P(3) = {}_6C_3(0.41)^3(0.59)^3 = 0.283$
$P(4) = {}_6C_4(0.41)^4(0.59)^2 = 0.148$
$P(5) = {}_6C_5(0.41)^5(0.59)^1 = 0.041$
$P(6) = {}_6C_6(0.41)^6(0.59)^0 = 0.004$

b.

x	$P(x)$
0	0.042
1	0.176
2	0.306
3	0.283
4	0.148
5	0.041
6	0.004

c.

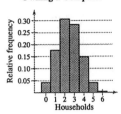

Owning a Computer

8a. Success: Selecting a clear day

$n = 31, p = 0.38, q = 0.62$

b. $\mu = np = (31)(0.38) = 11.78$

c. $\sigma^2 = npq = (31)(0.38)(0.62) = 7.3036$

d. $\sigma = \sqrt{\sigma^2} = 2.703$

e. On average, there are about 12 clear days during the year. The standard deviation is about 3 days.

4.2 EXERCISE SOLUTIONS

1. (a) $p = 0.50$ (graph is symmetric)

(b) $p = 0.20$ (graph is skewed right $\rightarrow p < 0.5$)

(c) $p = 0.80$ (graph is skewed left $\rightarrow p > 0.5$)

3. Is a binomial experiment.
Success: baby recovers
$n = 5, p = 0.80, q = 0.20, x = 0, 1, 2, . . . , 5$

5. Is not a binomial experiment because there are more than 2 possible outcomes for each trial.

7. $n = 5, p = .25$
(a) $P(3) \approx 0.088$
(b) $P(x \geq 3) = P(3) + P(4) + P(5) = 0.088 + 0.015 + .001 = 0.104$
(c) $P(x < 3) = 1 - P(x \geq 3) = 1 - 0.104 = 0.896$

9. $n = 10, p = 0.54$ (using binomial formula)
(a) $P(8) = 0.069$
(b) $P(x \geq 8) = P(8) + P(9) + P(10) = 0.069 + 0.018 + 0.002 = 0.089$
(c) $P(x < 8) = 1 - P(x \geq 8) = 1 - 0.089 = 0.911$

11. $n = 10, p = 0.21$ (using binomial formula)
(a) $P(2) \approx 0.301$
(b) $P(x \geq 2) = 1 - P(0) - P(1) = 1 - 0.095 - 0.252 = 0.653$
(c) $P(x < 2) = 1 - P(x \geq 2) = 1 - 0.653 = 0.347$

13. (a) $n = 6, p = 0.36$ (b)   **Basketball Fans**

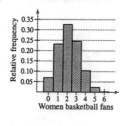

x	P(x)
0	0.069
1	0.232
2	0.326
3	0.245
4	0.103
5	0.023
6	0.002

(c) $\mu = np = (6)(0.36) = 2.16$ (d) $\sigma^2 = npq = (6)(0.36)(0.64) \approx 1.382$

(e) $\sigma = \sqrt{\sigma^2} \approx 1.176$

(f) On average 2.16, out of 6, women would consider themselves basketball fans. The standard deviation is 1.176 women.

$X = 0, 5,$ or 6 would be uncommon due to their low probabilities.

15. (a) $n = 4, p = 0.05$ (b)   **Donating Blood**

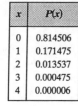
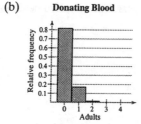

x	P(x)
0	0.814506
1	0.171475
2	0.013537
3	0.000475
4	0.000006

(c) $\mu = np = (4)(0.05) = 0.2$ (d) $\sigma^2 = npq = (4)(0.05)(0.95) = 0.19$

(e) $\sigma = \sqrt{\sigma^2} \approx 0.436$

(f) On average 0.2 eligible adults, out of every 4, give blood. The standard deviation is 0.436 adults.

$X = 2, 3,$ or 4 would be uncommon due to their low probabilities.

17. $n = 7, p = 0.10$

x	P(x)
0	0.478
1	0.372
2	0.124
3	0.023
4	0.003
5	0.000
6	0.000
7	0.000

19. $P(5, 2, 2, 1) = \dfrac{10!}{5!2!2!1!}\left(\dfrac{9}{16}\right)^5\left(\dfrac{3}{16}\right)^2\left(\dfrac{3}{16}\right)^2\left(\dfrac{1}{16}\right)^1 \approx 0.033$

4.3 MORE DISCRETE PROBABILITY DISTRIBUTIONS

4.3 Try It Yourself Solutions

1a. $P(1) = (0.23)(0.77)^0 = 0.23$

$P(2) = (0.23)(0.77)^1 = 0.177$

$P(3) = (0.23)(0.77)^2 = 0.136$

b. $P(x < 4) = P(1) + P(2) + P(3) = 0.543$

c. The probability that your first sale will occur before your fourth sales call is 0.543.

2a. $P(0) = \dfrac{3^0(2.71828)^{-3}}{0!} \approx 0.050$

$P(1) = \dfrac{3^1(2.71828)^{-3}}{1!} \approx 0.149$

$P(2) = \dfrac{3^2(2.71828)^{-3}}{2!} \approx 0.224$

$P(3) = \dfrac{3^3(2.71828)^{-3}}{3!} \approx 0.224$

$P(4) = \dfrac{3^4(2.71828)^{-3}}{4!} \approx 0.168$

3a. $\mu = \dfrac{2000}{20,000} = 0.10$

b. $\mu = 0.10, x = 3$

c. $P(3) = 0.0002$

4.3 EXERCISE SOLUTIONS

1. Geometric. We are interested in counting the number of trials until the first success.

3. Poisson. We are interested in counting the number of occurrences that take place within a given unit of space.

5. $p = 0.19$
  (a) $P(5) = (0.19)(0.81)^4 \approx 0.082$

  (b) $P(\text{sale on 1st, 2nd, or 3rd call}) = P(1) + P(2) + P(3) =$
  $$(0.19)(0.81)^0 + (0.19)(0.81)^1 + (0.19)(0.81)^2 \approx 0.469$$

  (c) $P(x > 3) = 1 - P(x \le 3) = 1 - 0.469 = 0.531$

7. $p = 0.535$

  (a) $P(2) = (0.535)(0.465)^1 \approx 0.249$

  (b) $P(\text{makes 1st or 2nd shot}) = P(1) + P(2) = (0.535)(0.465)^0 + (0.535)(0.465)^1 \approx 0.784$

  (c) (Binomial: $n = 2, p = .535$)
  $$P(0) = \frac{2!}{0!2!}(0.535)^0(0.465)^2 \approx 0.216$$

9. $\mu = 8$
  (a) $P(4) = \dfrac{8^4 e^{-8}}{4!} \approx 0.057$

  (b) $P(x \ge 4) = 1 - (P(0) + P(1) + P(2) + P(3))$
  $$\approx 1 - (0.0003 + 0.0027 + 0.0107 + 0.0286)$$
  $$= 0.9577$$

  (c) $P(x > 4) = 1 - (P(0) + P(1) + P(2) + P(3) + P(4))$
  $$\approx 1 - (0.0003 + 0.0027 + 0.0107 + 0.0286 + 0.0573)$$
  $$= 0.9004$$

11. $\mu = 0.6$

    (a) $P(1) = 0.3293$

    (b) $P(X \le 1) = P(0) + P(1) = 0.5488 + 0.3293 = 0.8781$

    (c) $P(X > 1) = 1 - P(X \le 1) = 1 - 0.8781 = 0.1219$

13. $p = 0.001$

    (a) $\mu = \dfrac{1}{p} = \dfrac{1}{0.001} = 1000$

$$\sigma^2 = \frac{q}{p^2} = \frac{0.999}{(0.001)^2} = 999{,}000$$

$$\sigma = \sqrt{\sigma^2} \approx 999.50$$

On average you would have to play 1000 times until you won the lottery. The standard deviation is 999.50.

    (b) 1000 times

        Lose money. On average you would win \$500 every 1000 times you play the lottery. Hence, the net gain would be $-\$500$

15. $\mu = 3.9$

    (a) $\sigma^2 = 3.9$

$$\sigma = \sqrt{\sigma^2} \approx 2.0$$

        The standard deviation is 2.0 strokes.

    (b) $P(X > 5) = 1 - P(X \le 5) = 1 - 0.801 = 0.199$

17. (a) $P(0) = \dfrac{_2C_0 \, _{13}C_3}{_{15}C_3} = \dfrac{(1)(286)}{(455)} \approx 0.629$

    (b) $P(1) = \dfrac{_2C_1 \, _{13}C_2}{_{15}C_3} = \dfrac{(2)(78)}{(455)} \approx 0.343$

    (c) $P(2) = \dfrac{_2C_2 \, _{13}C_1}{_{15}C_3} = \dfrac{(1)(13)}{(455)} \approx 0.029$

CHAPTER 4 REVIEW EXERCISE SOLUTIONS

1. Discrete

3. Continuous

5. No, $\Sigma P(x) \ne 1$

7. Yes

9. (a)

x	Frequency	$P(x)$
2	3	0.005
3	12	0.018
4	72	0.111
5	115	0.177
6	169	0.260
7	120	0.185
8	83	0.128
9	48	0.074
10	22	0.034
11	6	0.009
	650	1

(b)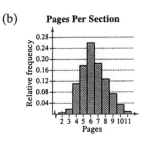

Pages Per Section

(c) $\mu = \Sigma x P(x) \approx 6.377$
$\sigma^2 = \Sigma(x - \mu)^2 P(x) \approx 2.858$
$\sigma = \sqrt{\sigma^2} \approx 1.691$

11. (a)

x	Frequency	$P(x)$
0	3	0.015
1	38	0.190
2	83	0.415
3	52	0.260
4	18	0.090
5	5	0.025
6	1	0.005
	200	1

(b)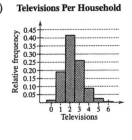

Televisions Per Household

(c) $\mu = \Sigma x P(x) \approx 2.315$
$\sigma^2 = \Sigma(x - \mu)^2 P(x) \approx 1.076$
$\sigma = \sqrt{\sigma^2} \approx 1.037$

13. $E(x) = \mu = \Sigma x P(x) = 3.37$

15. Yes, $n = 12, p = 0.30, q = 0.70, x = 0, 1, \ldots, 12$

17. $n = 8, p = 0.25$

(a) $P(3) = 0.208$

(b) $P(x \geq 3) = 1 - P(x < 3) = 1 - [P(0) + P(1) + P(2)] \approx 1 - [0.100 + 0.267 + 0.311] = 0.322$

(c) $P(x > 3) = 1 - P(x \leq 3) = 1 - [P(0) + P(1) + P(2) + P(3)]$
$= 1 - [0.100 + 0.267 + 0.311 + 0.208] = 0.114$

19. $n = 6, p = 0.43$ (use binomial formula)

(a) $P(3) = 0.294$

(b) $P(x \geq 3) = 1 - P(x < 3) = 1 - [P(0) + P(1) + P(2)] = 1 - [0.034 + 0.155 + 0.293] = 0.518$

(c) $P(3 \text{ or more}) = P(x \geq 3) = 0.518$

21. (a)

x	$P(x)$
0	0.006
1	0.050
2	0.167
3	0.294
4	0.293
5	0.155
6	0.034

(b)

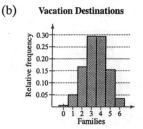

Vacation Destinations

(c) $\mu = np = (6)(0.57) = 3.42$
$\sigma^2 = npq = (6)(0.57)(0.43) = 1.4706$
$\sigma = \sqrt{\sigma^2} \approx 1.213$

(d) $P(x \le 3) = P(1) + P(2) + P(3) = 0.816$

23. $p = 0.167$

(a) $P(4) \approx 0.096$

(b) $P(x \le 4) = P(1) + P(2) + P(3) + P(4) \approx 0.518$

(c) $P(x > 3) = 1 - P(x \le 3) = 1 - [P(1) + P(2) + P(3)] \approx 0.606$

25. $\mu = \dfrac{7741}{42} \approx 184.310$ lightning deaths/year $\rightarrow \mu = \dfrac{184.310}{365} \approx 0.505$ deaths/day

(a) $P(0) = \dfrac{0.505^0 e^{-0.505}}{0!} \approx 0.604$

(b) $P(1) = \dfrac{0.505^1 e^{-0.505}}{1!} \approx 0.305$

(c) $P(x > 1) = 1 - [P(0) + P(1)] = 0.091$

CHAPTER 4 QUIZ SOLUTIONS

1. (a) Discrete because the random variable is countable.

(b) Continuous because the random variable has an infinite number of possible outcomes and cannot be counted.

2. (a)

x	Freq	$P(x)$
1	57	0.361
2	37	0.234
3	47	0.297
4	15	0.095
5	2	0.013
	158	1

(b)

Hurricane Intensity

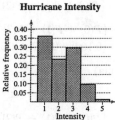

(c) $\mu = \Sigma x P(x) \approx 2.165$
$\sigma^2 = \Sigma(x - \mu)^2 P(x) \approx 1.125$
$\sigma = \sqrt{\sigma^2} \approx 1.061$

On average the intensity of a hurricane will be 2.165. The standard deviation is 1.061.

(d) $P(x \ge 4) = P(4) + P(5) = 0.095 + 0.013 = 0.108$

3. $n = 8, p = 0.80$

(a)

x	P(x)
0	0.000003
1	0.000082
2	0.001147
3	0.009175
4	0.045875
5	0.146801
6	0.293601
7	0.335544
8	0.167772

(b)

Successful Surgeries

(c) $\mu = np = (8)(0.80) = 6.4$
$\sigma^2 = npq = (8)(0.80)(0.20) = 1.28$
$\sigma = \sqrt{\sigma^2} \approx 1.131$

(d) $P(6) = 0.294$

(e) $P(x < 6) = 1 - P(x \geq 6) = 1 - [P(6) + P(7) + P(8)] = 1 - [0.294 + 0.336 + 0.168] = 0.202$

4. $\mu = 5$

(a) $P(5) = 0.1755$

(b) $P(x < 5) = P(0) + P(1) + P(2) + P(3) + P(4)$
$= 0.0067 + 0.0337 + 0.0842 + 0.1404 + 0.1755 = 0.4405$

(c) $P(0) = 0.0067$

Normal Probability Distributions

5.1 INTRODUCTION TO NORMAL DISTRIBUTIONS

5.1 Try It Yourself Solutions

1a. A: 45, B: 60, C: 45 (B has the greatest mean)

b. Curve C is more spread out so, curve C has the greatest standard deviation.

2a. Mean = 3.5 feet

b. Inflection points: 3.3 and 3.7
Standard deviation = 0.2 feet

3a. 85 is 1 standard deviation below the mean and 145 is 3 standard deviations above the mean.

b. 0.8385

5.1 EXERCISE SOLUTIONS

1. Answers will vary.

3. Answers will vary.
Similarities: Both curves will have the same line of symmetry.
Differences: One curve will be more spread out than the other.

5. No, the graph crosses the x-axis.

7. Yes, the graph fulfills the properties of the normal distribution.

9. 2 standard deviations

11. 95%: $(\mu \pm 2\sigma) \rightarrow (9, 21)$

13. $P(5.25 < x < 8.75) = P(\mu - \sigma < x < \mu + \sigma) \approx 0.68$

15. Curve B since the points of inflection are located at 2.98 and 3.00 (i.e. one standard deviation from the mean)

17. $P(2.94 < x < 3.06) = P(\mu - 3\sigma < x < \mu + 3\sigma) \approx 0.997$

19. 95%: $(\mu + 2\sigma) \rightarrow (19.86, 20.14)$

21. (a)

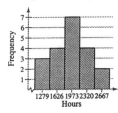

Light Bulb Lifespans

It is reasonable to assume that the lifespan is normally distributed since the histogram is nearly symmetric and bell-shaped.

(b) $\bar{x} = 1941.35 \; s \approx 432.385$

(c) The sample mean of 1941.35 hours is less than the claimed mean, so on the average the bulbs in the sample lasted for a shorter time. The sample standard deviation of 432 hours is greater than the claimed standard deviation, so the bulbs in the sample had a greater variation in lifespan than the manufacturer's claim.

23. $P(1970 < x < 2030) = P(\mu - \sigma < x < \mu + \sigma) \approx 0.68$

25. (a) $P(x < 5.3) = P(x < \mu - \sigma) \approx 0.16$

$(2000)(0.16) = 320$ students

(b) $P(5.3 < x < 7.1) = P(\mu - \sigma < x < \mu + \sigma) \approx 0.68$

$(2000)(0.68) = 1360$ students

(c) $P(x > 7.1) = P(x > \mu + \sigma) \approx 0.16$

$(2000)(0.16) = 320$ students

27.

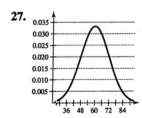

The normal distribution curve is centered at its mean (60) and has 2 points of inflection (48 and 72) representing $\mu \pm \sigma$.

29. (a) Area under curve = area of rectangle = (base)(height) = (1)(1) = 1

(b) $P(0.25 < x < 0.5) = $ (base)(height) = (0.25)(1) = 0.25

(c) $P(0.3 < x < 0.7) = $ (base)(height) = (0.4)(1) = 0.4

5.2 THE STANDARD NORMAL DISTRIBUTION

5.2 Try It Yourself Solutions

1a. $\mu = \$70, \sigma = \$8$

b. \$60: $z = \dfrac{60 - 70}{8} = -1.25$

\$71: $z = \dfrac{71 - 70}{8} = 0.125$

\$92: $z = \dfrac{92 - 70}{8} = 2.75$

c. A bill of \$60 is 1.25 standard deviations below the mean, a bill of \$71 is 0.125 standard deviations above the mean, and a bill of \$92 is 2.75 standard deviations above the mean.

2a. $\mu = \$70, \sigma = \$8$

b. $z = -0.75: x = 70 + (-0.75)(8) = 64$

$z = 4.29: x = 70 + (4.29)(8) = 104.32$

$z = -1.82: x = 70 + (-1.82)(8) = 55.44$

c. \$64 dollars is below the mean, \$104.32 is above the mean, and \$55.44 is below the mean.

3a. 0.0143

  b. 2.17

4a. 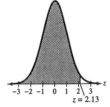   **b.** 0.9834

5a. 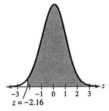   **b.** 0.0154

  c. Area $= 1 - 0.0154 = 0.9846$

6a. 0.0885

  b. 0.0154

  c. Area $= 0.0885 - 0.0154 = 0.0731$

5.2 EXERCISE SOLUTIONS

1. $\mu = 0, \sigma = 1$

3. "The" standard normal distribution is used to describe one specific normal distribution ($\mu = 0, \sigma = 1$). "A" normal distribution is used to describe a normal distribution with any mean and standard deviation.

5. (a) $x = 33{,}000$: $z = \dfrac{33{,}000 - 30{,}000}{2500} = 1.2$

      $x = 24{,}750$: $z = \dfrac{24{,}750 - 30{,}000}{2500} = -2.1$

      $x = 30{,}000$: $z = \dfrac{30{,}000 - 30{,}000}{2500} = 0$

  (b) lower 10% $\rightarrow z \approx -1.28$

      $x = \mu + z\sigma = 30{,}000 + (-1.28)(2500) = 26{,}800$

      Advertise a tire life of 26,800 miles.

7. (a) $x = 85$: $z = \dfrac{85 - 76}{7} = 1.29$

      $x = 91$: $z = \dfrac{91 - 76}{7} = 2.14$

      $x = 70$: $z = \dfrac{70 - 76}{7} \approx -0.86$

      $x = 67$: $z = \dfrac{67 - 76}{7} \approx -1.29$

  (b) The scores seem typical because all are within 3 standard deviations of the mean. There are an equal number of scores above and below the mean so the scores do not appear to be either above or below average.

9. (a) $x = \mu + Z\sigma = 153 + (1.2)(12) = 167.4$

(b) $x = \mu + Z\sigma = 153 + (-2.4)(12) = 124.2$

11. $z = -0.67, 0, 0.67$

13. 0.33

15. 1.29

17. (Area left of $z = 1.2$) − (Area left of $z = 0$) = 0.8849 − 0.5 = 0.3849

19. (Area left of $z = 1.5$) − (Area left of $z = -0.5$) = 0.9332 − 0.3085 = 0.6247

21. 0.9382

23. $1 - 0.1711 = 0.8289$

25. 0.005

27. $1 - 0.95 = 0.05$

29. $0.975 - 0.5 = 0.475$

31. $0.5 - 0.0630 = 0.437$

33. $0.8810 - 0.3300 = 0.551$

35. $0.0015 + 0.0485 = 0.05$

37. $P(z < 1.45) = 0.9265$

39. $P(z > -1.95) = 1 - P(z < -1.95) = 1 - 0.0256 = 0.9744$

41. $P(z < -0.55) = 0.2912$

43. $P(z > 1.05) = 1 - P(z < 1.05) = 1 - 0.8531 = 0.1469$

45. $P(0 < z < 2.05) = 0.9798 - 0.5 = 0.4798$

47. $P(-0.89 < z < 0) = 0.5 - 0.1867 = 0.3133$

49. $P(-0.95 < z < 1.44) = 0.9251 - 0.1711 = 0.7540$

51. $P(z < -2.58 \text{ or } z > 2.58) = 2(0.0049 = 0.0098$

53. $P(-2 < z < 2) = 0.9772 - 0.0228 = 0.9544$

5.3 APPLICATIONS OF NORMAL DISTRIBUTIONS

5.3 Try It Yourself Solutions

1a. GRE: $z = \dfrac{x - \mu}{\sigma} = \dfrac{1775 - 1500}{300} \approx 0.92$

   MAT: $z = \dfrac{x - \mu}{\sigma} = \dfrac{54 - 50}{5} \approx 0.8$

b. GRE: 0.8212
   MAT: 0.7881

c. GRE score is better.

2a.

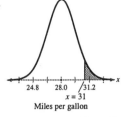

Miles per gallon

b. $z = \dfrac{x - \mu}{\sigma} = \dfrac{31 - 28}{1.6} = 1.875$

c. $P(z < 1.88) = 0.9696$
$P(z > 1.88) = 1 - 0.9696 = 0.0304$

d. The probability that a randomly selected Escort will get more than 31 mpg is 0.0304.

3a.

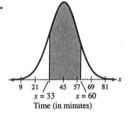

Time (in minutes)

b. $z = \dfrac{x - \mu}{\sigma} = \dfrac{33 - 45}{12} = -1$

$z = \dfrac{x - \mu}{\sigma} = \dfrac{60 - 45}{12} = 1.25$

c. $P(z < -1) = 0.1587$
$P(z < 1.25) = 0.8944$

d. $P(-1 < z < 1.25) = 0.8944 - 0.1587 = 0.7357$

4a. Read user's guide for the technology tool.

b. Enter the data.

c. $P(190 < x < 225) = P(-1 < z < 0.4) = 0.4967$

5a.

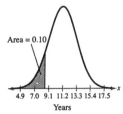

Years

b. $z = -1.28$

c. $x = \mu + z\sigma = 11.2 + (-1.28)(2.1) = 8.512$

d. The maximum length of time an employee could have worked and still be laid off is 8.512 years.

5.3 EXERCISE SOLUTIONS

1. SAT: $z = \dfrac{x - \mu}{\sigma} = \dfrac{1115 - 1000}{200} \approx 0.58$

   ACT: $z = \dfrac{x - \mu}{\sigma} = \dfrac{23 - 20}{5} = 0.6$

   ACT score was better.

3. (a) $P(x < 66) = P(z < -1.10) = 0.1357$

   (b) $P(66 < x < 72) = P(-1.10 < z < 0.97) = 0.8340 - 0.1357 = 0.6983$

   (c) $P(x > 72) = P(z > 0.97) = 1 - P(z < 0.97) = 1 - 0.8340 = 0.1660$

5. (a) $P(x < 20) = P(z < -1.02) = 0.1539$

   (b) $P(20 < x < 29) = P(-1.02 < z < 1.12) = 0.8686 - 0.1539 = 0.7147$

   (c) $P(x > 29) = P(z > 1.12) = 1 - P(z < 1.12) = 1 - 0.8686 = 0.1314$

7. (a) $P(x < 2.5) = P(z < -2.5) = 0.0062$

   (b) $P(2.5 < x < 7.5) = P(-2.5 < z < 2.5) = 0.9938 - 0.0062 = 0.9876$

   (c) $P(x > 7.5) = P(z > 2.5) = 1 - P(z < 2.5) = 1 - 0.9938 = 0.0062$

9. (a) $P(x > 75) = P(z > 2) = 1 - P(z < 2) = 1 - 0.9772 = 0.0228 \rightarrow 2.28\%$

   (b) $P(x < 72) = P(z < 0.97) = 0.8340$
   $(100)(0.8340) = 83.4$

   (c) 90th percentile $\rightarrow z \approx 1.28$
   $x = \mu + z\sigma = 69.2 + (1.28)(2.9) = 72.912$

   (d) 1st quartile $\rightarrow z \approx -0.67$
   $x = \mu + z\sigma = 69.2 + (-0.67)(2.9) = 67.257$

11. (a) $P(x > 25) = P(z > 0.17) = 1 - P(z < 0.17) = 1 - 0.5675 = 0.4324 \rightarrow 43.24\%$

    (b) $P(x < 18) = P(z < -1.5) = 0.0668$
    $(150)(0.0668) = 10.02$

    (c) 95th percentile $\rightarrow z \approx 1.645$
    $x = \mu + z\sigma = 24.3 + (1.645)(4.2) = 31.209$

    (d) 1st quartile $\rightarrow z \approx -0.67$
    $x = \mu + z\sigma = 24.3 + (-0.67)(4.2) = 21.486$

13. (a) $P(x > 2) = P(z > -3) = 1 - P(z < -3) = 1 - 0.0013 = 0.9987 \rightarrow 99.87\%$

    (b) $P(x < 3) = P(z < -2) = 0.0228$
    $(35)(0.0228) = 0.798$

    (c) top 25% $\rightarrow$ 75th percentile $\rightarrow z \approx 0.67$
    $x = \mu + z\sigma = 5 + (0.67)(1) = 5.67$

    (d) bottom 15% $\rightarrow$ 15th percentile $\rightarrow z \approx -1.04$
    $x = \mu + z\sigma = 5 + (-1.04)(1) = 3.96$

15. (a) bottom 10% $\rightarrow$ 10th percentile $\rightarrow z \approx -1.28$
    $x = \mu + z\sigma = 10.2 + (-1.28)(1.7) = 8.024$

(b) bottom $7\% \rightarrow$ 7th percentile $\rightarrow z \approx -1.48$

$x = \mu + z\sigma = 10.2 + (-1.48)(1.7) = 7.684$

17. Out of control, since the 10th observation plotted beyond 3 standard deviations.

19. Out of control, since the first nine observations lie below the mean.

5.4 Try It Yourself Solutions

1a.

Sample	Mean	Sample	Mean	Sample	Mean
1,1	1	3,1	2	6,1	3.5
1,2	1.5	3,2	2.5	6,2	4
1,3	2	3,3	3	6,3	4.5
1,5	3	3,5	4	6,5	5.5
1,6	3.5	3,6	4.5	6,6	6
1,7	4	3,7	5	6,7	6.5
2,1	1.5	5,1	3	7,1	4
2,2	2	5,2	3.5	7,2	4.5
2,3	2.5	5,3	4	7,3	5
2,5	3.5	5,5	5	7,5	6
2,6	4	5,6	5.5	7,6	6.5
2,7	4.5	5,7	6	7,7	7

b. $\mu_{\bar{x}} = 4, \sigma_{\bar{x}}^2 \approx 2.33, \sigma_{\bar{x}} \approx 1.53$

c. $\mu = 4, \sigma^2 = 4.68, \sigma = 2.16$

2a. $\mu_{\bar{x}} = \mu = 64, \sigma_{\bar{x}} = \dfrac{\sigma}{\sqrt{n}} = \dfrac{9}{\sqrt{100}} = 0.9$

b. $n = 100$

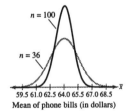

59.5 61.0 62.5 64.0 65.5 67.0 68.5
Mean of phone bills (in dollars)

3a. $\mu_{\bar{x}} = \mu = 3.5, \sigma_{\bar{x}} = \dfrac{\sigma}{\sqrt{n}} = \dfrac{0.2}{\sqrt{16}} = 0.05$

b.

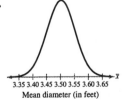

3.35 3.40 3.45 3.50 3.55 3.60 3.65
Mean diameter (in feet)

4a. $\mu_{\bar{x}} = \mu = 33, \sigma_{\bar{x}} = \dfrac{\sigma}{\sqrt{n}} = \dfrac{4}{\sqrt{45}} \approx 0.596$

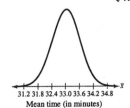

31.2 31.8 32.4 33.0 33.6 34.2 34.8
Mean time (in minutes)

b. $\bar{x} = 27$: $z = \dfrac{\bar{x} - \mu}{\dfrac{\sigma}{\sqrt{n}}} = \dfrac{27 - 33}{\dfrac{4}{\sqrt{45}}} = -10.06$

$\bar{x} = 35$: $z = \dfrac{\bar{x} - \mu}{\dfrac{\sigma}{\sqrt{n}}} = \dfrac{35 - 33}{\dfrac{4}{\sqrt{45}}} = 3.35$

c. $P(z < -10.06) \approx 0$

$P(z < 3.35) = 0.9996$

d. $P(27 < \bar{x} < 35) = P(-10.06 < z < 3.35) = 0.9996 - 0 = 0.9996$

5a. $\mu_{\bar{x}} = \mu = 125{,}700, \sigma_{\bar{x}} = \dfrac{\sigma}{\sqrt{n}} = \dfrac{26{,}000}{\sqrt{16}} = 6500$

112,700 125,700 138,700
Mean sales price (in dollars)

b. $\bar{x} = 80{,}000$: $z = \dfrac{\bar{x} - \mu}{\dfrac{\sigma}{\sqrt{n}}} = \dfrac{80{,}000 - 125{,}700}{\dfrac{26{,}000}{\sqrt{16}}} = -7.03$

c. $P(\bar{x} > 80{,}000) = P(z > -7.03) = 1 - P(z < -7.03) \approx 1 - 0 = 1$

6a. $x = 700$: $z = \dfrac{x - \mu}{\sigma} = \dfrac{700 - 625}{150} = 0.5$

$\bar{x} = 700$: $z = \dfrac{\bar{x} - \mu}{\dfrac{\sigma}{\sqrt{n}}} = \dfrac{700 - 625}{\dfrac{150}{\sqrt{10}}} = 1.58$

b. $P(z < 0.5) = 0.6915$

$P(z < 1.58) = 0.9429$

c. There is a 69% chance an <u>individual receiver</u> will cost less than $700. There is a 94% chance that the <u>mean of a sample of 10 receivers</u> is less than $700.

5.4 EXERCISE SOLUTIONS

1. False, the sampling distribution of sample means for sample size 40 has a mean equal to the population mean and a standard deviation equal to the population standard deviation divided by the square root of 40.

3. {000, 002, 004, 006, 008, 020, 022, 024, 026, 028, 040, 042, 044, 046, 048, 060, 062, 064, 066, 068, 080, 082, 084, 086, 088, 200, 202, 204, 206, 208, 220, 222, 224, 226, 228, 240, 242, 244, 246, 248, 260, 262, 264, 266, 268, 280, 282, 284, 286, 288, 400, 402, 404, 406, 408, 420, 422, 424, 426, 428, 440, 442, 444, 446, 448, 460, 462, 464, 466, 468, 480, 482, 484, 486, 488, 600, 602, 604, 606, 608, 620, 622, 624, 626, 628, 640, 642, 644, 646, 648, 660, 662, 664, 666, 668, 680, 682, 684, 686, 688, 800, 802, 804, 806, 808, 820, 822, 824, 826, 828, 840, 842, 844, 846, 848, 860, 862, 864, 866, 868, 880, 882, 884, 886, 888}

$\mu_{\bar{x}} = 4, \sigma_{\bar{x}} = 1.633$

$\mu = 4, \sigma = 2.828$

5. $\mu_{\bar{x}} = 87.5, \sigma_{\bar{x}} = \dfrac{\sigma}{\sqrt{n}} = \dfrac{6.25}{\sqrt{12}} \approx 1.804$

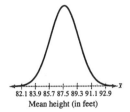

82.1 83.9 85.7 87.5 89.3 91.1 92.9
Mean height (in feet)

7. $\mu_{\bar{x}} = 114.7, \sigma_{\bar{x}} = \dfrac{\sigma}{\sqrt{n}} = \dfrac{38}{\sqrt{20}} \approx 8.497$

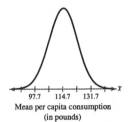

97.7   114.7   131.7
Mean per capita consumption
(in pounds)

9. $\mu_{\bar{x}} = 87.5, \sigma_{\bar{x}} = \dfrac{\sigma}{\sqrt{n}} = \dfrac{6.25}{\sqrt{24}} \approx 1.276$

$\mu_{\bar{x}} = 87.5, \sigma_{\bar{x}} = \dfrac{\sigma}{\sqrt{n}} = \dfrac{6.25}{\sqrt{36}} \approx 1.042$

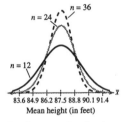

$n = 36$
$n = 24$
$n = 12$
83.6 84.9 86.2 87.5 88.8 90.1 91.4
Mean height (in feet)

As the sample size increases, the standard error decreases.

11. (c) since $\mu_{\bar{x}} = 16.5, \sigma_{\bar{x}} = \dfrac{\sigma}{\sqrt{n}} = \dfrac{11.9}{\sqrt{100}} = 1.19$

13. $z = \dfrac{\bar{x} - \mu}{\dfrac{\sigma}{\sqrt{n}}} = \dfrac{29,500 - 28,000}{\dfrac{1700}{\sqrt{35}}} = 5.22$

$P(\bar{x} < 29,500) = P(z < 5.22) \approx 1$

15. $z = \dfrac{\bar{x} - \mu}{\dfrac{\sigma}{\sqrt{n}}} = \dfrac{1.075 - 1.080}{\dfrac{0.045}{\sqrt{32}}} \approx 0.63$

$z = \dfrac{\bar{x} - \mu}{\dfrac{\sigma}{\sqrt{n}}} = \dfrac{1.090 - 1.080}{\dfrac{0.045}{\sqrt{32}}} \approx 1.26$

17. $z = \dfrac{\bar{x} - \mu}{\dfrac{\sigma}{\sqrt{n}}} = \dfrac{68 - 64}{\dfrac{2.75}{\sqrt{60}}} \approx 11.27$

$P(\bar{x} > 68) = P(z > 11.27) = 0$

19. $z = \dfrac{\bar{x} - \mu}{\sigma} = \dfrac{70 - 64}{2.75} \approx 2.18$

$P(x < 70) = P(z < 2.18) = 0.9854$

$z = \dfrac{\bar{x} - \mu}{\dfrac{\sigma}{\sqrt{n}}} = \dfrac{70 - 64}{\dfrac{2.75}{\sqrt{20}}} \approx 9.76$

$P(\bar{x} < 70) = P(z < 9.76) \approx 1$

It is more likely to select a sample of 20 women with a mean height less than 70 inches.

21. $z = \dfrac{\bar{x} - \mu}{\dfrac{\sigma}{\sqrt{n}}} = \dfrac{127.9 - 128}{\dfrac{0.20}{\sqrt{40}}} \approx 3.16$

$P(\bar{x} < 127.9) = P(z < -3.16) = 0.0008$

Yes, it is very unlikely that we would have randomly sampled 40 cans with a mean equal to 127.9 ounces.

23. Use the finite correction factor since $n = 55 > 40 = 0.05N$.

$z = \dfrac{\bar{x} - \mu}{\dfrac{\sigma}{\sqrt{n}}\sqrt{\dfrac{N-n}{N-1}}} = \dfrac{1.005 - 1.007}{\dfrac{0.009}{\sqrt{55}}\sqrt{\dfrac{800-55}{800-1}}} \approx -1.71$

$P(\bar{x} < 1.005) = P(z < -1.71) = 0.0436$

25. Use the finite correction factor since $n = 30 > 25 = 0.05N$.

$z = \dfrac{\bar{x} - \mu}{\dfrac{\sigma}{\sqrt{n}}\sqrt{\dfrac{N-n}{N-1}}} = \dfrac{2.5 - 3.32}{\dfrac{1.09}{\sqrt{30}}\sqrt{\dfrac{500-30}{500-1}}} \approx -4.25$

$z = \dfrac{\bar{x} - \mu}{\dfrac{\sigma}{\sqrt{n}}\sqrt{\dfrac{N-n}{N-1}}} = \dfrac{4 - 3.32}{\dfrac{1.09}{\sqrt{30}}\sqrt{\dfrac{500-30}{500-1}}} \approx 3.52$

$P(2.5 < \bar{x} < 4) = P(-4.25 < z < 3.52) \approx 1 - 0 = 1$

5.5 NORMAL APPROXIMATIONS TO BINOMIAL DISTRIBUTIONS

5.5 Try It Yourself Solutions

1a. $n = 70, p = 0.08, q = 0.92$

b. $np = 5.6, nq = 64.4$

c. Since $np \geq 5$ and $nq \geq 5$, the normal distribution can be used.

d. $\mu = np = (70)(0.08) = 5.6$
$\sigma = \sqrt{npq} = \sqrt{(70)(0.08)(0.92)} \approx 2.270$

2a. (1) $57, 58, \ldots, 83$ (2) $\ldots, 52, 53, 54$

  b. (1) $56.5 < x < 83.5$ (2) $x < 54.5$

3a. $x > 10.5$

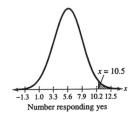

  b. $z = \dfrac{x - \mu}{\sigma} = \dfrac{10.5 - 5.6}{2.270} \approx 2.16$

  c. $P(z < 2.16) = 0.9846$

  d. $P(x > 10.5) = P(z > 2.16) = 1 - P(z < 2.16) = 0.0154$

  e. The probability that more than 10 respond yes is 0.0154.

4a. $x < 65.5$

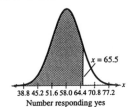

  b. $z = \dfrac{x - \mu}{\sigma} = \dfrac{65.5 - 58}{6.42} \approx 1.17$

  c. $P(x < 65.5) = P(z < 1.17) = 0.8790$

  d. The probability that at most 65 people will say yes is 0.8790.

5a. $60.5 < x < 61.5$

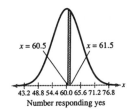

  b. $z = \dfrac{x - \mu}{\sigma} = \dfrac{60.5 - 60}{5.59} \approx 0.09$

     $z = \dfrac{x - \mu}{\sigma} = \dfrac{61.5 - 60}{5.59} \approx 0.27$

  c. $P(z < 0.09) = 0.5359$
     $P(z < 0.27) = 0.6064$

  d. $P(0.09 < z < 0.27) = 0.6064 - 0.5359 = 0.0705$

  e. The probability that exactly 61 people will respond yes is 0.0705.

5.5 EXERCISE SOLUTIONS

1. $n = 100, p = 0.70, q = 0.30$
$np = 70 \geq 5, nq = 30 \geq 5$
Use normal distribution.
$\mu = np = (100)(0.70) = 70$
$\sigma = \sqrt{npq} = \sqrt{(100)(0.70)(0.30)} \approx 4.583$

3. $n = 10, p = 0.89, q = 0.11$
$np = 8.9 \geq 5, nq = 1.1 < 5$
Cannot use normal distribution.

5. d

7. a

9. a

11. c

13. Binomial: $P(5 \leq x \leq 7) = 0.162 + 0.198 + 0.189 = 0.549$
Normal: $P(4.5 \leq x \leq 7.5) = P(-0.97 < z < 0.56) = 0.7123 - 0.1660 = 0.5463$

15. $n = 30, p = 0.07 \rightarrow np = 2.1$ and $nq = 27.9$
Cannot use normal distribution.

(a) $P(x = 10) = {}_{30}C_{10}(0.07)^{10}(0.93)^{20} \approx 0.0000199$

(b) $P(x \geq 10) = 1 - P(x < 10)$
$= 1 - [{}_{30}C_0(0.07)^0(0.93)^{30} + {}_{30}C_1(0.07)^1(0.93)^{29} + \cdots + {}_{30}C_9(0.07)^9(0.93)^{21}]$
$= 1 - .999977 \approx 0.000023$

(c) $P(x < 10) \approx 0.999977$ (see part b)

(d) $n = 100, p = 0.07 \rightarrow np = 7$ and $nq = 93$
Use normal distribution.
$z = \dfrac{x - \mu}{\sigma} = \dfrac{4.5 - 7}{2.55} \approx -0.98$

$P(x < 5) = P(x < 4.5) = P(z < -0.98) = 0.1635$

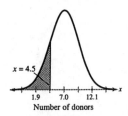

$x = 4.5$
1.9   7.0   12.1
Number of donors

17. $n = 40, p = 0.52 \rightarrow np = 20.8$ and $nq = 19.2$
Use the normal distribution.

(a) $z = \dfrac{x - \mu}{\sigma} = \dfrac{15.5 - 20.8}{3.160} \approx -1.68$
$P(x \leq 15) = P(x < 15.5) = P(z < -1.68) = 0.0465$

Favor chocolate chip cookies

(b) $z = \dfrac{x - \mu}{\sigma} = \dfrac{14.5 - 20.8}{3.160} \approx 1.99$

$P(x \geq 15) = P(x > 14.5) = P(z > -1.99) = 1 - P(z < 1.99) = 1 - 0.0233 = 0.9767$

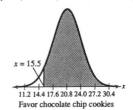

Favor chocolate chip cookies

(c) $P(x > 15) = P(x > 15.5) = P(z > -1.68) = 1 - P(z < -1.68) = 1 - 0.0465 = 0.9535$

Favor chocolate chip cookies

(d) $n = 650, p = 0.52 \rightarrow np = 338$ and $nq = 312$

Use normal distribution.

$z = \dfrac{x - \mu}{\sigma} = \dfrac{350.5 - 338}{12.74} \approx 0.98$

$P(x > 350) = P1(x > 350.5) = P(z > 0.98) = 1 - P(z < 0.98) = 1 - 0.8365 = 0.1635$

299 312 325 338 351 364 377
Favor chocolate chip cookies

19. $n = 250, p = 0.70$

60% say no $\rightarrow 250(0.6) = 150$ say no while 100 say yes.

$z = \dfrac{x - \mu}{\sigma} = \dfrac{99.5 - 175}{7.25} = -10.41$

$P(\text{less than } 100 \text{ yes}) = P(x < 100) = P(x < 99.5) = P(z < -10.41) \approx 0$

It is highly unlikely that 60% responded no.

21. $n = 100, p = 0.75$

$z = \dfrac{x - \mu}{\sigma} = \dfrac{69.5 - 75}{4.33} \approx 1.27$

$P(\text{reject claim}) = P(x < 70) = P(x < 69.5) = P(z < -1.27) = 0.1020$

CHAPTER 5 REVIEW EXERCISE SOLUTIONS

1. $\mu = 15, \sigma = 3$

3. $95\% \to \mu \pm 2\sigma \to 670 \pm 2(65) \to (540, 800)$

5. $P(605 < x < 735) = P(\mu - \sigma < x < \mu + \sigma) = 0.68$

7. $x = 1.32$: $z = \dfrac{x - \mu}{\sigma} = \dfrac{1.32 - 1.5}{0.08} = -2.25$

   $x = 1.54$: $z = \dfrac{x - \mu}{\sigma} = \dfrac{1.54 - 1.5}{0.08} = 0.5$

   $x = 1.66$: $z = \dfrac{x - \mu}{\sigma} = \dfrac{1.66 - 1.5}{0.08} = 2$

   $x = 1.78$: $z = \dfrac{x - \mu}{\sigma} = \dfrac{1.78 - 1.5}{0.08} = 3.5$

9. 0.2005

11. 0.3936

13. $1 - 0.9535 = 0.0465$

15. $0.5 - 0.0505 = 0.4495$

17. $0.9115 - 0.5596 = 0.35199$

19. $0.0668 + 0.0668 = 0.1336$

21. $P(z < 1.28) = 0.8997$

23. $P(-2.15 < x < 1.55) = 0.9394 - 0.0158 = 0.9236$

25. $P(z < -2.50 \text{ or } z > 2.50) = 2(0.0062) = 0.0124$

27. $x = 99$: $z = \dfrac{x - \mu}{\sigma} = \dfrac{99 - 122}{14} = -1.64$

   $x = 70$: $z = \dfrac{x - \mu}{\sigma} = \dfrac{70 - 83}{9} = -1.44$

   The first participant had the lower reading.

29. (a) $z = \dfrac{x - \mu}{\sigma} = \dfrac{1900 - 2200}{625} \approx -0.48$

   $P(x < 1900) = P(z < -0.48) = 0.3156$

   (b) $z = \dfrac{x - \mu}{\sigma} = \dfrac{2000 - 2200}{625} \approx -0.32$

   $z = \dfrac{x - \mu}{\sigma} = \dfrac{25900 - 2200}{625} \approx 0.48$

   $P(2000 < x < 2500) = P1(-0.32 < z < 0.48) = 0.6844 - 0.3745 = 0.3099$

   (c) $z = \dfrac{x - \mu}{\sigma} = \dfrac{2450 - 2200}{625} = -0.4$

   $P(x > 2450) = P1(z > 0.4) = 0.3446$

31. (a) top $15\% \to z \approx 1.04$

   $\mu + z\sigma = 1.4 + (1.04)(0.20) = 1.608$ meters

   Body length would have to be more than 1.608 meters.

(b) top $5\% \to z \approx 1.645$

$\mu + z\sigma = 1.4 + (1.645)(0.20) = 1.729$ meters

Body length would have to be more than 1.729 meters.

33. {0 0 0, 0 0 200, 0 0 40, 0 0 600, 0 0 80, 0 200 0, 0 200 200, 0 200 40, 0 200 600, 0 200 80, 0 40 0, 0 40 200, 0 40 40, 0 40 600, 0 40 80, 0 600 0, 0 600 200, 0 600 40, 0 600 600, 0 600 80, 0 80 0, 0 80 200, 0 80 40, 0 80 600, 0 80 80, 200 0 0, 200 0 200, 200 0 40, 200 0 600, 200 0 80, 200 200 0, 200 200 200, 200 200 40, 200 200 600, 200 200 80, 200 40 0, 200 40 200, 200 40 40, 200 40 600, 200 40 80, 200 600 0, 200 600 200, 200 600 40, 200 600 600, 200 600 80, 200 80 0, 200 80 200, 200 80 40, 200 80 600, 200 80 80, 40 0 0, 40 0 200, 40 0 40, 40 0 600, 40 0 80, 40 200 0, 40 200 200, 40 200 40, 40 200 600, 40 200 80, 40 40 0, 40 40 200, 40 40 40, 40 40 600, 40 40 80, 40 600 0, 40 600 200, 40 600 40, 40 600 600, 40 600 80, 40 80 0, 40 80 200, 40 80 40, 40 80 600, 40 80 80, 600 0 0, 600 0 200, 600 0 40, 600 0 600, 600 0 80, 600 200 0, 600 200 200, 600 200 40, 600 200 600, 600 200 80, 600 40 0, 600 40 200, 600 40 40, 600 40 600, 600 40 80, 600 600 0, 600 600 200, 600 600 40, 600 600 600, 600 600 80, 600 80 0, 600 80 200, 600 80 40, 600 80 600, 600 80 80, 80 0 0, 80 0 200, 80 0 40, 80 0 600, 80 0 80, 80 200 0, 80 200 200, 80 200 40, 80 200 600, 80 200 80, 80 40 0, 80 40 200, 80 40 40, 80 40 600, 80 40 80, 80 600 0, 80 600 200, 80 600 40, 80 600 600, 80 600 80, 80 80 0, 80 80 200, 80 80 40, 80 80 600, 80 80 80}

$\mu = 184, \sigma \approx 218.504$

$\mu_{\bar{x}} = 184, \sigma_{\bar{x}} \approx 126.153$

35. (a) $z = \dfrac{\bar{x} - \mu}{\dfrac{\sigma}{\sqrt{n}}} = \dfrac{1900 - 2200}{\dfrac{625}{\sqrt{12}}} \approx -1.66$

$P(\bar{x} < 1900) = P(z < -1.66) = 0.0485$

(b) $z = \dfrac{\bar{x} - \mu}{\dfrac{\sigma}{\sqrt{n}}} = \dfrac{2000 - 2200}{\dfrac{625}{\sqrt{12}}} \approx -1.11$

$z = \dfrac{\bar{x} - \mu}{\dfrac{\sigma}{\sqrt{n}}} = \dfrac{2500 - 2200}{\dfrac{625}{\sqrt{12}}} \approx 1.66$

$P(2000 < \bar{x} < 2500) = P(-1.11 < z < 1.66) = 0.9515 - 0.1335 = 0.8180$

(c) $z = \dfrac{\bar{x} - \mu}{\dfrac{\sigma}{\sqrt{n}}} = \dfrac{2450 - 2200}{\dfrac{625}{\sqrt{12}}} \approx 1.39$

$P(\bar{x} > 2450) = P(z > 1.39) = 0.0823$

37. $\mu_{\bar{x}} = 154.8, \sigma_{\bar{x}} \dfrac{\sigma}{\sqrt{n}} = \dfrac{51.6}{\sqrt{35}} \approx 8.72$

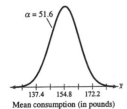

$\alpha = 51.6$

137.4   154.8   172.2

Mean consumption (in pounds)

39. (a) $z = \dfrac{\bar{x} - \mu}{\dfrac{\sigma}{\sqrt{n}}} = \dfrac{20{,}000 - 21{,}000}{\dfrac{1500}{\sqrt{45}}} \approx -4.47$

$P(\bar{x} < 20{,}000) = P(z < -4.47) \approx 0$

(b) $z = \dfrac{\bar{x} - \mu}{\dfrac{\sigma}{\sqrt{n}}} = \dfrac{22{,}500 - 21{,}000}{\dfrac{1500}{\sqrt{45}}} \approx 6.71$

$P(\bar{x} > 22{,}500) = P(z > 6.71) \approx 0$

41. $n = 12, p = 0.59, q = 0.41$

$np = 7.08 > 5$, but $nq = 4.92 < 5$

Cannot use the normal distribution.

43. $P(x \geq 25) = P(x > 24.5)$

45. $n = 45, p = 0.65 \rightarrow np = 29.25, nq = 15.75$

Use normal distribution.

$\mu = np = 29.25, \sigma = \sqrt{npq} = \sqrt{45(0.65)(0.35)} \approx 3.200$

$z = \dfrac{x - \mu}{\sigma} = \dfrac{20.5 - 29.25}{3.200} \approx -2.73$

$P(x \leq 20) = P(x < 20.5) = P(z < -2.73) = 0.0032$

$x = 20.5$

22.85   29.25   35.65

Children saying yes

CHAPTER 5 QUIZ SOLUTIONS

1. (a) $P(z > -2.10) = 0.9821$

(b) $P(z < 3.22) = 0.9994$

(c) $P(-2.33 < z < 2.33) = 0.9901 - 0.0099 = 0.9802$

(d) $P(z < -1.75 \text{ or } z > -0.75) = 0.0401 + 0.7734 = 0.8135$

2. (a) $z = \dfrac{x - \mu}{\sigma} = \dfrac{5.36 - 5.5}{0.08} \approx -1.75$

$z = \dfrac{x - \mu}{\sigma} = \dfrac{5.64 - 5.5}{0.08} \approx 1.75$

$P(5.36 < x < 5.64) = P(-1.75 < z < 1.75) = 0.9599 - 0.0401 = 0.9198$

(b) $z = \dfrac{x - \mu}{\sigma} = \dfrac{-5.00 - (-8.2)}{7.84} \approx 0.41$

$z = \dfrac{x - \mu}{\sigma} = \dfrac{0 - (-8.2)}{7.84} \approx 1.05$

$P(-5.00 < x < 0) = P(0.41 < z < 1.05) = 0.8531 - 0.6591 = 0.1940$

(c) $z = \dfrac{x - \mu}{\sigma} = \dfrac{0 - 18.5}{9.25} = -2$

$z = \dfrac{x - \mu}{\sigma} = \dfrac{37 - 18.5}{9.25} = 2$

$P(x < 0 \text{ or } x > 37) = P(z < -2 \text{ or } z > 2) = 2(0.0228) = 0.0456$

3. $z = \dfrac{x - \mu}{\sigma} = \dfrac{315 - 276.1}{34.4} \approx 1.13$

$P(x > 315) = P(z > 1.13) = 0.1292$

4. $z = \dfrac{x - \mu}{\sigma} = \dfrac{250 - 276.1}{34.4} \approx -0.76$

$z = \dfrac{x - \mu}{\sigma} = \dfrac{305 - 276.1}{34.4} \approx 0.84$

$P(250 < x < 305) = P(-0.76 < z < 0.84) = 0.7995 - 0.2236 = 0.5759$

5. $P(x > 250) = P(z > -0.76) = 0.7764 \rightarrow 77.64\%$

6. $z = \dfrac{x - \mu}{\sigma} = \dfrac{300 - 276.1}{34.4} \approx 0.69$

$P(x < 300) = P(z < 0.69) = 0.7549$

$(2000)(0.7549) = 1509.8$

7. top $5\% \rightarrow z \approx 1.645$

$\mu + z\sigma = 276.1 + (1.645)(34.4) = 332.688$

8. bottom $25\% \rightarrow z \approx -0.67$

$\mu + z\sigma = 276.1 + (-0.67)(34.4) = 253.052$

9. $z = \dfrac{\bar{x} - \mu}{\dfrac{\sigma}{\sqrt{n}}} = \dfrac{300 - 276.1}{\dfrac{34.4}{\sqrt{60}}} \approx 5.38$

$P(\bar{x} > 300) = P(z > 5.38) \approx 0$

10. $z = \dfrac{x - \mu}{\sigma} = \dfrac{300 - 276.1}{34.4} \approx 0.69$

$P(\bar{x} > 300) = P(z > 0.69) = 0.2451$

$z = \dfrac{\bar{x} - \mu}{\dfrac{\sigma}{\sqrt{n}}} = \dfrac{305 - 276.1}{\dfrac{34.4}{\sqrt{15}}} \approx 3.25$

$P(\bar{x} > 300) = P(z > 3.25) = 0.0006$

You are more likely to select one student with a test score greater than 300.

11. $n = 24, p = 0.68 \rightarrow np = 16.32, nq = 7.68$

Use normal distribution.

$\mu = np = 16.32 \qquad \sigma = \sqrt{npq} \approx 2.285$

12. $z = \dfrac{x - \mu}{\sigma} = \dfrac{15.5 - 16.32}{2.285} \approx -0.36$

$P(x \le 15) = P(x < 15.5) = P(z < -0.36) = 0.3594$

Confidence Intervals

6.1 Try It Yourself Solutions

Section 6.1

1a. $\bar{x} \approx 14.767$

b. The mean number of sentences per magazine advertisement is 14.767.

2a. $Z_c = 1.96, n = 30, s = 16.536$

b. $E = Z_c \dfrac{s}{\sqrt{n}} = 1.96 \dfrac{16.536}{\sqrt{30}} \approx 5.917$

c. You are 95% confident that the maximum error of the estimate is about 5.917 sentences per magazine advertisement.

3a. $\bar{x} \approx 14.767, E \approx 5.917$

b. $\bar{x} - E = 14.767 - 5.917 = 8.850$
$\bar{x} + E = 14.767 + 5.917 = 20.684$

c. You are 95% confident that the mean number of sentences per magazine advertisements is between 8.850 and 20.684.

4. 75% CI: (11.293, 18.240)
85% CI: (10.421, 19.113)

5a. $n = 20, \bar{x} = 22.9, \sigma = 1.5, Z_c = 1.282$

b. $E = Z_c \dfrac{\sigma}{\sqrt{n}} = 1.282 \dfrac{1.5}{\sqrt{20}} \approx 0.430$

c. $\bar{x} - E = 22.9 - 0.430 = 22.47$
$\bar{x} + E = 22.9 + 0.430 = 23.33$

d. You are 80% confident that the mean age of the students is between 22.47 and 23.33 years.

6a. $Z_c = 1.96, E = 2, s \approx 5.0$

b. $n = \left(\dfrac{Z_c s}{E}\right)^2 = \left(\dfrac{1.96 \cdot 5.0}{2}\right)^2 = 24.01 \rightarrow 25$

c. You should have at least 25 magazine advertisements in your sample.

6.1 EXERCISE SOLUTIONS

1. You are more likely to be correct using an interval estimate since it is unlikely that a point estimate will equal the population mean exactly.

3. d; As the level of confidence increases, z_c increases therefore creating wider intervals.

5. 1.28

7. $|\bar{x} - \mu| = |3.8 - 4.27| = 0.47$

9. $|\bar{x} - \mu| = |26.43 - 24.67| = 1.76$

11. $E = z_c \dfrac{s}{\sqrt{n}} = 1.645 \dfrac{2.5}{\sqrt{36}} \approx 0.685$

13. $\bar{x} \pm z_c \dfrac{s}{\sqrt{n}} = 15.2 \pm 1.645 \dfrac{2.0}{\sqrt{60}} \approx (14.775, 15.625)$

15. $\bar{x} \pm z_c \dfrac{s}{\sqrt{n}} = 4.27 \pm 1.96 \dfrac{0.3}{\sqrt{42}} \approx (4.179, 4.361)$

17. 90% CI: $\bar{x} \pm z_c \dfrac{s}{\sqrt{n}} = 280.90 \pm 1.645 \dfrac{123.70}{\sqrt{32}} \approx (244.928, 316.872)$

    95% CI: $\bar{x} \pm z_c \dfrac{s}{\sqrt{n}} = 280.90 \pm 1.96 \dfrac{123.70}{\sqrt{32}} \approx (238.040, 323.760)$

19. 90% CI: $\bar{x} \pm z_c \dfrac{s}{\sqrt{n}} = 26.8 \pm 1.645 \dfrac{8.0}{\sqrt{156}} \approx (25.746, 27.854)$

    95% CI: $\bar{x} \pm z_c \dfrac{s}{\sqrt{n}} = 26.8 \pm 1.96 \dfrac{8.0}{\sqrt{156}} \approx (25.545, 28.055)$

21. $\bar{x} \pm z_c \dfrac{s}{\sqrt{n}} = 100 \pm 1.96 \dfrac{17.50}{\sqrt{40}} \approx (94.577, 105.423)$

23. $\bar{x} \pm z_c \dfrac{s}{\sqrt{n}} = 100 \pm 1.96 \dfrac{17.50}{\sqrt{80}} \approx (96.165, 103.835)$

$n = 40$ CI is wider because we have taken a smaller sample giving us less information about the population.

25. $\bar{x} \pm z_c \dfrac{s}{\sqrt{n}} = 10.452 \pm 2.575 \dfrac{2.130}{\sqrt{56}} \approx (9.719, 11.185)$

27. $\bar{x} \pm z_c \dfrac{s}{\sqrt{n}} = 10.452 \pm 2.575 \dfrac{5.130}{\sqrt{56}} \approx (8.687, 12.217)$

$s = 5.130$ CI is wider because of the increased variability within the population.

29. (a) An increase in the level of confidence will widen the confidence interval.

    (b) An increase in the sample size will narrow the confidence interval.

    (c) An increase in the standard deviation will widen the confidence interval.

31. $\bar{x} = \dfrac{\Sigma x}{n} = \dfrac{136}{15} \approx 9.067$ 90% CI: $\bar{x} \pm z_c \dfrac{\sigma}{\sqrt{n}} = 9.067 \pm 1.645 \dfrac{1.5}{\sqrt{15}} \approx (8.430, 9.704)$

    99% CI: $\bar{x} \pm z_c \dfrac{\sigma}{\sqrt{n}} = 9.067 \pm 2.575 \dfrac{1.5}{\sqrt{15}} \approx (8.070, 10.064)$

    99% CI is wider.

33. $n \left(\dfrac{z_c \sigma}{E} \right)^2 = \left(\dfrac{1.96 \cdot 4.8}{1} \right)^2 \approx 88.510 \rightarrow 89$

35. (a) $n \left(\dfrac{z_c \sigma}{E} \right)^2 = \left(\dfrac{1.96 \cdot 2.8}{0.5} \right)^2 \approx 120.473 \rightarrow 121$

    (b) $n \left(\dfrac{z_c \sigma}{E} \right)^2 = \left(\dfrac{2.575 \cdot 2.8}{0.5} \right)^2 \approx 207.936 \rightarrow 208$

99% CI requires larger sample because more information is needed from the population to be 99% confident.

37. (a) $n\left(\dfrac{z_c\sigma}{E}\right)^2 = \left(\dfrac{1.645 \cdot 0.85}{0.25}\right)^2 \approx 31.282 \rightarrow 32$

(b) $n\left(\dfrac{z_c\sigma}{E}\right)^2 = \left(\dfrac{1.645 \cdot 0.85}{0.15}\right)^2 \approx 86.893 \rightarrow 87$

$E = 0.15$ requires a larger sample size. As the error size decreases, a larger sample must be taken to obtain enough information from the population to ensure desired accuracy.

39. (a) $n\left(\dfrac{z_c\sigma}{E}\right)^2 = \left(\dfrac{2.575 \cdot 0.25}{0.1}\right)^2 \approx 41.441 \rightarrow 42$

(b) $n\left(\dfrac{z_c\sigma}{E}\right)^2 = \left(\dfrac{2.575 \cdot 0.30}{0.1}\right)^2 \approx 59.676 \rightarrow 60$

$\sigma = 0.30$ requires a larger sample size. Due to the increased variability in the population, a larger sample size is needed to ensure the desired accuracy.

41. (a) An increase in the level of confidence will increase the minimum sample size required.

(b) An increase (larger E) in the error tolerance will decrease the minimum sample size required.

(c) An increase in the population standard deviation will increase the minimum sample size required.

43. $\bar{x} = 307.375, s \approx 11.190, n = 32$

$\bar{x} \pm z_c\dfrac{s}{\sqrt{n}} = 307.375 \pm 1.96\dfrac{11.190}{\sqrt{32}} \approx (303.498, 311.252)$

45. $\bar{x} = 14.440, s \approx 2.124, n = 30$

$\bar{x} \pm z_c\dfrac{s}{\sqrt{n}} = 14.440 \pm 1.96\dfrac{2.124}{\sqrt{30}} \approx (13.680, 15.200)$

47. (a) $\sqrt{\dfrac{N-n}{N-1}} = \sqrt{\dfrac{1000-500}{1000-1}} \approx 0.707$

(b) $\sqrt{\dfrac{N-n}{N-1}} = \sqrt{\dfrac{1000-100}{1000-1}} \approx 0.949$

(c) $\sqrt{\dfrac{N-n}{N-1}} = \sqrt{\dfrac{1000-75}{1000-1}} \approx 0.962$

(d) $\sqrt{\dfrac{N-n}{N-1}} = \sqrt{\dfrac{1000-50}{1000-1}} \approx 0.975$

(e) The finite population correction factor approaches 1 as the sample size decreases while the population size remains the same.

49. $n = \left(\dfrac{z_c\sigma}{E}\right)^2 \rightarrow \sqrt{n} = \dfrac{z_c\sigma}{E} \rightarrow E = \dfrac{z_c\sigma}{\sqrt{n}}$

6.2 CONFIDENCE INTERVALS FOR THE MEAN (SMALL SAMPLES)

6.2 Try It Yourself Solutions

1a. d.f. $= n - 1 = 22 - 1 = 21$

b. $c = 0.90$

c. 1.721

2a. 90% CI: $t_c = 1.753$

$$E = t_c \frac{s}{\sqrt{n}} = 1.753 \frac{10}{\sqrt{16}} \approx 4.383$$

99% CI: $t_c = 2.947$

$$E = t_c \frac{s}{\sqrt{n}} = 2.947 \frac{10}{\sqrt{16}} \approx 7.368$$

b. 90% CI: $\bar{x} \pm E = 162 \pm 4.383 \approx (157.618, 166.383)$
99% CI: $\bar{x} \pm E = 162 \pm 7.368 \approx (154.633, 169.368)$

c. You are 90% confident that the mean temperature of coffee sold is between 157.618° and 166.383°.
You are 99% confident that the mean temperature of coffee sold is between 154.633° and 169.368°.

3a. 90% CI: $t_c = 1.729$

$$E = t_c \frac{s}{\sqrt{n}} = 1.729 \frac{0.42}{\sqrt{20}} \approx 0.162$$

95% CI: $t_c = 2.093$

$$E = t_c \frac{s}{\sqrt{n}} = 2.093 \frac{0.42}{\sqrt{20}} \approx 0.197$$

b. 90% CI: $\bar{x} \pm E = 6.93 \pm 0.162 \approx (6.768, 7.092)$
95% CI: $\bar{x} \pm E = 6.93 \pm 0.197 \approx (6.733, 7.127)$

c. You are 90% confident that the mean mortgage interest rate is contained between 6.768% and 7.092%.
You are 99% confident that the mean mortgage interest rate is contained between 6.733% and 7.127%.

4a. Is $n \geq 30$? No
Is the population normally distributed? Yes
Is σ known? No
Use the t-distribution to construct the 90% CI.

6.2 EXERCISE SOLUTIONS

1. 1.833

3. 2.947

5. (a) $E = z_c \frac{s}{\sqrt{n}} = 1.96 \frac{5}{\sqrt{16}} \approx 2.450$

  (b) $E = t_c \frac{s}{\sqrt{n}} = 2.131 \frac{5}{\sqrt{16}} \approx 2.664$

7. (a) $\bar{x} \pm t_c \frac{s}{\sqrt{n}} = 12.5 \pm 2.015 \frac{2.0}{\sqrt{6}} \approx (10.855, 14.145)$

  (b) $\bar{x} \pm t_c \frac{s}{\sqrt{n}} = 12.5 \pm 1.645 \frac{2.0}{\sqrt{6}} \approx (11.157, 13.843)$

   t-CI is wider.

9. (a) $\bar{x} \pm t_c \dfrac{s}{\sqrt{n}} = 4.3 \pm 2.650 \dfrac{0.34}{\sqrt{14}} \approx (4.059, 4.541)$

(b) $\bar{x} \pm z_c \dfrac{s}{\sqrt{n}} = 4.3 \pm 2.326 \dfrac{0.34}{\sqrt{14}} \approx (4.089, 4.511)$

t-CI is wider.

11. $\bar{x} \pm t_c \dfrac{s}{\sqrt{n}} = 75 \pm 2.776 \dfrac{12.50}{\sqrt{5}} \approx (59.482, 90.518)$

$E = t_c \dfrac{s}{\sqrt{n}} = 2.776 \dfrac{12.50}{\sqrt{5}} \approx 15.518$

13. $\bar{x} \pm z_c \dfrac{\sigma}{\sqrt{n}} = 75 \pm 1.96 \dfrac{15}{\sqrt{5}} \approx (61.852, 88.148)$

$E = z_c \dfrac{\sigma}{\sqrt{n}} = 1.96 \dfrac{15}{\sqrt{5}} \approx 13.148$

t-CI is wider.

15. (a) $\bar{x} \pm t_c \dfrac{s}{\sqrt{n}} = 4.3 \pm 1.833 \dfrac{1.2}{\sqrt{10}} \approx (3.604, 4.996)$

(b) $\bar{x} \pm z_c \dfrac{s}{\sqrt{n}} = 4.3 \pm 1.645 \dfrac{1.2}{\sqrt{500}} \approx (4.212, 4.338)$

t-CI is wider.

17. (a) $\bar{x} = 2174.75$

(b) $s \approx 100.341$

(c) $\bar{x} \pm t_c \dfrac{s}{\sqrt{n}} = 2174.75 \pm 3.250 \dfrac{100.341}{\sqrt{10}} \approx (2071.626, 2277.874)$

19. (a) $\bar{x} \approx 909.083$

(b) $s \approx 305.266$

(c) $\bar{x} \pm t_c \dfrac{s}{\sqrt{n}} = 909.083 \pm 3.106 \dfrac{305.266}{\sqrt{12}} \approx (635.374, 1182.792)$

21. $n \geq 30 \rightarrow$ use normal distribution

$\bar{x} \pm z_c \dfrac{s}{\sqrt{n}} = 1.25 \pm 1.96 \dfrac{0.01}{\sqrt{70}} \approx (1.248, 1.252)$

23. $\bar{x} = 24, s = 3, n < 30, \sigma$ known, and pop normally distributed $\rightarrow$ use t-distribution

$\bar{x} \pm t_c \dfrac{s}{\sqrt{n}} = 24 \pm 2.064 \dfrac{3}{\sqrt{25}} \approx (22.762, 25.238)$

25. $n < 30, \sigma$ unknown, and pop <u>not</u> normally distributed $\rightarrow$ cannot use either the normal or t-distributions.

27. $n = 25, \bar{x} = 56.0, s = 0.25$

$\pm t_{0.99} \rightarrow$ 99% t-CI

$\bar{x} \pm t_c \dfrac{s}{\sqrt{n}} = 56.0 \pm 2.797 \dfrac{0.25}{\sqrt{25}} \approx (55.860, 56.140)$

They are not making good tennis balls since desired bounce height of 55.5 inches is not contained between 55.850 and 56.140 inches.

6.3 CONFIDENCE INTERVALS FOR POPULATION PROPORTIONS

6.3 Try It Yourself Solutions

1a. $x = 98, n = 1470$

 b. $\hat{p} = \dfrac{98}{1470} \approx 0.067$

2a. $\hat{p} \approx 0.067, \hat{q} \approx 0.933$

 b. $z_c = 1.645$

 $$E = z_c \sqrt{\dfrac{\hat{p}\hat{q}}{n}} = 1.645 \quad \sqrt{\dfrac{0.067 \cdot 0.933}{1470}} \approx 0.011$$

 c. $\hat{p} \pm E = 0.067 \pm 0.011 \approx (0.056, 0.078)$

 d. You are 90% confident that the proportion of adults that admired Abraham Lincoln the most is contained between 5.6% and 7.8%.

3a. $n = 935, \hat{p} \approx 0.16$

 b. $\hat{q} = 1 - \hat{p} = 1 - 0.16 \approx 0.84$

 c. $n\hat{p} = 935 \cdot 0.16 \approx 149.600 \geq 5$
 $n\hat{q} = 935 \cdot 0.84 \approx 785.400 \geq 5$
 Distribution of $\hat{p}$ is approximately normal.

 d. $z_c = 2.575$

 e. $\hat{p} \pm z_c \sqrt{\dfrac{\hat{p}\hat{q}}{n}} = 0.16 \pm 2.575 \sqrt{\dfrac{0.16 \cdot 0.84}{935}} \approx (0.129, 0.191)$

 f. You are 99% confident that the proportion of adults who think that trains are the safest mode of transportation is contained between 12.9% and 19.1%.

4a. $\hat{p} - 0.04, \hat{q} = 0.96$

 b. $z_c = 1.645, E = 0.05$

 c. $n = \hat{p}\hat{q}\left(\dfrac{z_c}{E}\right)^2 = 0.04 \cdot 0.96\left(\dfrac{1.645}{0.05}\right)^2 \approx 41.565 \rightarrow 42$

 d. At least 42 adults should be included in the sample.

6.3 EXERCISE SOLUTIONS

1. $\hat{p} = \dfrac{x}{n} = \dfrac{83}{1040} \approx 0.080$
 $\hat{q} = 1 - \hat{p} \approx 0.920$

3. $\hat{p} = \dfrac{x}{n} = \dfrac{4080}{34,000} \approx 0.120$
 $\hat{q} = 1 - \hat{p} \approx 0.880$

5. $\hat{p} = \dfrac{x}{n} = \dfrac{93}{1418} \approx 0.066$
 $\hat{q} = 1 - \hat{p} \approx 0.934$

7. $\hat{p} = \dfrac{x}{n} = \dfrac{181}{262} \approx 0.691$

   $\hat{q} = 1 - \hat{p} \approx 0.309$

9. 95% CI: $\hat{p} \pm z_c\sqrt{\dfrac{\hat{p}\hat{q}}{n}} = 0.080 \pm 1.96\sqrt{\dfrac{0.080 \cdot 0.920}{1040}} \approx (0.064, 0.096)$

   99% CI: $\hat{p} \pm z_c\sqrt{\dfrac{\hat{p}\hat{q}}{n}} = 0.080 \pm 2.575\sqrt{\dfrac{0.080 \cdot 0.920}{1040}} \approx (0.058, 0.102)$

   99% CI is wider.

11. 95% CI: $\hat{p} \pm z_c\sqrt{\dfrac{\hat{p}\hat{q}}{n}} = 0.120 \pm 1.96\sqrt{\dfrac{0.120 \cdot 0.880}{34,000}} \approx (0.117, 0.123)$

    99% CI: $\hat{p} \pm z_c\sqrt{\dfrac{\hat{p}\hat{q}}{n}} = 0.120 \pm 55\sqrt{\dfrac{0.120 \cdot 0.880}{34,000}} \approx (0.115, 0.125)$

    99% CI is wider.

13. 95% CI: $\hat{p} \pm z_c\sqrt{\dfrac{\hat{p}\hat{q}}{n}} = 0.066 \pm 1.96\sqrt{\dfrac{0.066 \cdot 0.934}{1418}} \approx (0.053, 0.079)$

    99% CI: $\hat{p} \pm z_c\sqrt{\dfrac{\hat{p}\hat{q}}{n}} = 0.066 \pm 2.575\sqrt{\dfrac{0.066 \cdot 0.934}{1418}} \approx (0.049, 0.083)$

    99% CI is wider.

15. 95% CI: $\hat{p} \pm z_c\sqrt{\dfrac{\hat{p}\hat{q}}{n}} = 0.691 \pm 1.96\sqrt{\dfrac{0.691 \cdot 0.309}{262}} \approx (0.635, 0.747)$

    99% CI: $\hat{p} \pm z_c\sqrt{\dfrac{\hat{p}\hat{q}}{n}} = 0.691 \pm 2.575\sqrt{\dfrac{0.691 \cdot 0.309}{262}} \approx (0.617, 0.765)$

    99% CI is wider.

17. (a) $n = \hat{p}\hat{q}\left(\dfrac{z_c}{E}\right)^2 = 0.5 \cdot 0.5\left(\dfrac{1.96}{0.03}\right)^2 \approx 1067.111 \rightarrow 1068$

    (b) $n = \hat{p}\hat{q}\left(\dfrac{z_c}{E}\right)^2 = 0.26 \cdot 0.74\left(\dfrac{1.96}{0.03}\right)^2 \approx 821.249 \rightarrow 822$

    (c) Having an estimate of the proportion reduces the minimum samples size needed.

19. (a) $n = \hat{p}\hat{q}\left(\dfrac{z_c}{E}\right)^2 = 0.5 \cdot 0.5\left(\dfrac{2.054}{0.025}\right)^2 \approx 1687.57 \rightarrow 1688$

    (b) $n = \hat{p}\hat{q}\left(\dfrac{z_c}{E}\right)^2 = 0.25 \cdot 0.75\left(\dfrac{2.054}{0.025}\right)^2 \approx 1265.67 \rightarrow 1266$

    (c) Having an estimate of the proportion reduces the minimum sample size needed.

21. (a) $\hat{p} = 0.61, n = 500$

    $\hat{p} \pm z_c\sqrt{\dfrac{\hat{p}\hat{q}}{n}} = 0.61 \pm 2.575\sqrt{\dfrac{0.61 \cdot 0.39}{500}} \approx (0.554, 0.666)$

    (b) $\hat{p} = 0.44, n = 500$

    $\hat{p} \pm z_c\sqrt{\dfrac{\hat{p}\hat{q}}{n}} = 0.44 \pm 2.575\sqrt{\dfrac{0.44 \cdot 0.56}{500}} \approx (0.383, 0.497)$

    It is unlikely that the two proportions are equal because the confidence intervals estimating the proportions do not overlap.

23. $31.4\% \pm 1\% \rightarrow (30.4\%, 32.4\%)$

$$E = z_c \sqrt{\frac{\hat{p}\hat{q}}{n}} \rightarrow z_c = E\sqrt{\frac{n}{\hat{p}\hat{q}}} = 0.01\sqrt{\frac{8451}{0.314 \cdot 0.686}} \approx 1.981 \rightarrow z_c = 1.98 \rightarrow c = 0.952$$

$(30.4\%, 32.4\%)$ is approximately a 95.2% CI.

25. If $n\hat{p} < 5$ or $n\hat{q} < 5$, the sampling distribution of $\hat{p}$ may not be normally distributed; therefore preventing the use of z_c when calculating the confidence interval.

27.

p	$q = 1 - p$	pq		p	$q = 1 - p$	pq
0.1	0.9	0.09		0.45	0.55	0.2475
0.2	0.8	0.16		0.46	0.54	0.2484
0.3	0.7	0.21		0.47	0.53	0.2491
0.4	0.6	0.24		0.48	0.52	0.2496
0.5	0.5	0.25		0.49	0.51	0.2499
0.6	0.4	0.24		0.50	0.50	0.2500
0.7	0.3	0.21		0.51	0.49	0.2499
0.8	0.2	0.16		0.52	0.48	0.2496
0.9	0.1	0.09		0.53	0.47	0.2491
1.0	0.0	0.00		0.54	0.46	0.2484
				0.55	0.45	0.2475

$\hat{p} = 0.5$ give the maximum value of $\hat{p}\hat{q}$.

6.4 CONFIDENCE INTERVALS FOR VARIANCE AND STANDARD DEVIATION

6.4 Try It Yourself Solutions

1a. d.f. $= n - 1 = 24$
level of confidence $= 0.95$

b. Area to the right of χ_R^2 is 0.025.
Area to the left of χ_L^2 is 0.975.

c. $\chi_R^2 = 39.364$, $\chi_L^2 = 12.401$

2a. 90% CI: $\chi_R^2 = 42.557$, $\chi_L^2 = 17.708$
95% CI: $\chi_R^2 = 45.722$, $\chi_L^2 = 16.047$

b. *90% CI for σ^2:* $\left(\frac{(n-1)s^2}{\chi_R^2}, \frac{(n-1)s^2}{\chi_L^2}\right) = \left(\frac{29 \cdot (1.2)^2}{42.557}, \frac{29 \cdot (1.2)^2}{17.708}\right) \approx (0.981, 2.358)$

95% CI for σ^2: $\left(\frac{(n-1)s^2}{\chi_R^2}, \frac{(n-1)s^2}{\chi_L^2}\right) = \left(\frac{29 \cdot (1.2)^2}{45.722}, \frac{29 \cdot (1.2)^2}{16.047}\right) \approx (0.913, 2.602)$

c. 90% CI for σ: $(\sqrt{0.981}, \sqrt{2.358}) = (0.990, 1.536)$
95% CI for σ: $(\sqrt{0.913}, \sqrt{2.602}) = (0.956, 1.613)$

d. You are 90% confident that the population variance is between 0.990 and 1.536. You are 95% confident that the population variance is between 0.956 and 1.613.

6.4 EXERCISE SOLUTIONS

1. $\chi_R^2 = 16.919$, $\chi_L^2 = 3.325$

3. $\chi_R^2 = 35.479$, $\chi_L^2 = 10.283$

5. $\chi_R^2 = 52.336$, $\chi_L^2 = 13.121$

7. (a) $s = 0.00843$

$$\left(\frac{(n-1)s^2}{\chi_R^2}, \frac{(n-1)s^2}{\chi_L^2}\right) = \left(\frac{13 \cdot (0.00843)^2}{22.362}, \frac{13 \cdot (0.00843)^2}{5.892}\right) \approx (0.0000413, 0.000157)$$

(b) $\left(\sqrt{0.0000413}, \sqrt{0.000157}\right) \approx (0.00643, 0.0125)$

9. (a) $s = 0.253$

$$\left(\frac{(n-1)s^2}{\chi_R^2}, \frac{(n-1)s^2}{\chi_L^2}\right) = \left(\frac{17 \cdot (0.253)^2}{35.718}, \frac{17 \cdot (0.253)^2}{5.697}\right) \approx (0.0305, 0.191)$$

(b) $\left(\sqrt{0.0305}, \sqrt{0.191}\right) \approx (0.175, 0.437)$

11. (a) $\left(\frac{(n-1)s^2}{\chi_R^2}, \frac{(n-1)s^2}{\chi_L^2}\right) = \left(\frac{11 \cdot (3.25)^2}{26.757}, \frac{11 \cdot (3.25)^2}{2.603}\right) \approx (4.342, 44.636)$

(b) $\left(\sqrt{4.342}, \sqrt{44.636}\right) \approx (2.084, 6.681)$

13. (a) $\left(\frac{(n-1)s^2}{\chi_R^2}, \frac{(n-1)s^2}{\chi_L^2}\right) = \left(\frac{9 \cdot (26)^2}{16.919}, \frac{9 \cdot (26)^2}{3.325}\right) \approx (359.596, 1829.774)$

(b) $\left(\sqrt{359.596}, \sqrt{1829.774}\right) \approx (18.963, 42.776)$

15. (a) $\left(\frac{(n-1)s^2}{\chi_R^2}, \frac{(n-1)s^2}{\chi_L^2}\right) = \left(\frac{19 \cdot (107)^2}{32.852}, \frac{19 \cdot (107)^2}{8.907}\right) \approx (6621.545, 24{,}422.477)$

(b) $\left(\sqrt{6621.545}, \sqrt{24{,}422.477}\right) \approx (81.373, 156.277)$

17. 90% CI for σ: $(0.00643, 0.0125)$

Yes, because the confidence interval is below 0.015.

CHAPTER 6 REVIEW EXERCISE SOLUTIONS

1. (a) $\bar{x} \approx 103.5$

(b) $s \approx 34.663$

$$E = z_c \frac{s}{\sqrt{n}} = 1.645 \frac{34.663}{\sqrt{40}} \approx 9.016$$

3. $\bar{x} \pm z_c \frac{s}{\sqrt{n}} = 10.3 \pm 1.96 \frac{0.277}{\sqrt{100}} \approx (10.246, 10.354)$

5. $s = 34.663$

$$n = \left(\frac{z_c \sigma}{E}\right)^2 = \left(\frac{1.96 \cdot 34.663}{10}\right)^2 \approx 46.158 \rightarrow 47$$

7. $t_c = 1.415$

9. $E = z_c \frac{s}{\sqrt{n}} = 1.645 \frac{23.4}{\sqrt{16}} \approx 9.623$

11. $\bar{x} \pm z_c\dfrac{s}{\sqrt{n}} = 52.8 \pm 1.645\dfrac{23.4}{\sqrt{16}} \approx (43.177, 62.423)$

13. $\bar{x} \pm t_c\dfrac{s}{\sqrt{n}} = 80 \pm 1.761\dfrac{14}{\sqrt{15}} \approx (73.634, 86.366)$

15. $\hat{p} = \dfrac{x}{n} = \dfrac{357}{850} = 0.420, \quad \hat{q} = 0.580$

17. $\hat{p} = \dfrac{x}{n} = \dfrac{61}{209} = 0.292, \quad \hat{q} = 0.708$

19. $\hat{p} \pm z_c\sqrt{\dfrac{\hat{p}\hat{q}}{n}} = 0.420 \pm 1.96\sqrt{\dfrac{0.420 \cdot 0.580}{850}} \approx (0.387, 0.453)$

21. $\hat{p} \pm z_c\sqrt{\dfrac{\hat{p}\hat{q}}{n}} = 0.292 \pm 1.645\sqrt{\dfrac{0.292 \cdot 0.708}{209}} \approx (0.240, 0.344)$

23. $n = \hat{p}\hat{q}\left(\dfrac{z_c}{E}\right)^2 = 0.23 \cdot 0.77\left(\dfrac{1.96}{0.05}\right)^2 \approx 272.139 \rightarrow 273$

25. $\chi_R^2 = 23.337, \chi_L^2 = 4.404$

27. $\chi_R^2 = 14.067, \chi_L^2 = 2.167$

29. $s = 0.073$

95% CI for σ^2: $\left(\dfrac{(n-1)s^2}{\chi_R^2}, \dfrac{(n-1)s^2}{\chi_L^2}\right) = \left(\dfrac{15 \cdot (0.073)^2}{27.488}, \dfrac{15 \cdot (0.073)^2}{6.262}\right) \approx (0.003, 0.013)$

95% CI for σ: $\left(\sqrt{0.003}, \sqrt{0.013}\right) \approx (0.055, 0.114)$

CHAPTER 6 QUIZ SOLUTIONS

1. (a) $\bar{x} \approx 100.057$

(b) $s \approx 26.286$

$$E = t_c\dfrac{s}{\sqrt{n}} = 2.069\dfrac{26.286}{\sqrt{24}} \approx 11.101$$

(c) $\bar{x} \pm t_c\dfrac{s}{\sqrt{n}} = 100.057 \pm 2.069\dfrac{26.286}{\sqrt{24}} \approx (88.956, 111.158)$

You are 95% confident that the population mean repair costs is contained between $88.96 and $111.16.

2. $n = \left(\dfrac{z_c\sigma}{E}\right)^2 = \left(\dfrac{2.575 \cdot 22.50}{10}\right) \approx 33.568 \rightarrow$ sample size must be 34

3. (a) $\bar{x} = 6.610$ (b) $s \approx 3.376$

(c) $\bar{x} \pm t_c\dfrac{s}{\sqrt{n}} = 6.610 \pm 1.833\dfrac{3.376}{\sqrt{10}} \approx (4.653, 8.567)$

(d) $\bar{x} \pm z_c\dfrac{\sigma}{\sqrt{n}} = 6.610 \pm 1.645\dfrac{3.5}{\sqrt{10}} \approx (4.789, 8.431)$

4. $\bar{x} \pm t_c\dfrac{s}{\sqrt{n}} = 3705 \pm 2.365\dfrac{566}{\sqrt{8}} \approx (3231.737, 4178.263)$

5. (a) $\hat{p} = \dfrac{x}{n} = \dfrac{1320}{2000} = 0.660$

(b) $\hat{p} \pm z_c \sqrt{\dfrac{\hat{p}\hat{q}}{n}} = 0.660 \pm 1.645 \sqrt{\dfrac{0.660 \cdot 0.340}{2000}} \approx (0.643, 0.677)$

(c) $n = \hat{p}\hat{q}\left(\dfrac{z_c}{E}\right)^2 = 0.660 \cdot 0.340 \left(\dfrac{2.575}{0.04}\right)^2 \approx 929.945$

6. (a) $\left(\dfrac{(n-1)s^2}{\chi_R^2}, \dfrac{(n-1)s^2}{\chi_L^2}\right) = \left(\dfrac{23 \cdot (26.286)^2}{38.076}, \dfrac{23 \cdot (26.286)^2}{11.689}\right) \approx (417.374, 1359.563)$

(b) $\left(\sqrt{417.374}, \sqrt{1359.563}\right) \approx (20.430, 36.872)$

CUMULATIVE TEST SOLUTIONS FOR CHAPTERS 4-6

1. $\hat{p} = \dfrac{x}{n} = \dfrac{365}{474} \approx 0.770$

$\hat{p} \pm z_c \sqrt{\dfrac{\hat{p}\hat{q}}{n}} = 0.770 \pm 1.96 \sqrt{\dfrac{0.770 \cdot 0.230}{474}} \approx (0.732, 0.808)$

2. $n = \hat{p}\hat{q}\left(\dfrac{z_c}{E}\right)^2 = 0.770 \cdot 0.230 \left(\dfrac{2.575}{0.02}\right)^2 \approx 2935.870 \rightarrow 2936$

3. $n = 12$, $p = 0.77$

$P(x \geq 10) = P(10) + P(11) + P(12)$

$\qquad = {}_{12}C_{11}(0.77)^{11}(0.23)^1 + {}_{12}C_{11}(0.77)^{11}(0.23)^1 + {}_{12}C_{11}(0.77)^{11}(0.23)^1$

$\qquad = 0.256 + 0.156 + 0.043 \approx 0.455$

4. $\mu = np = (474)(0.770) = 364.980$

$\sigma^2 = npq = (474)(0.770)(0.230) \approx 83.945$

$\sigma = \sqrt{npq} \approx 9.162$

You would expect 364.98 women to say that the media have a negative effect on women's health. The standard deviation is 9.162.

5. $np = 364.980 \geq 5$ and $nq = 109.02 \geq 5$

Use normal distribution.

$\mu = np = 364.980$

$\sigma = \sqrt{npq} \approx 9.162$

6. $\bar{x} \pm z_c \dfrac{s}{\sqrt{n}} = 25.6 \pm 1.645 \dfrac{3.5}{\sqrt{474}} \approx (25.336, 25.864)$

7. Normal distribution was used since $n \geq 30$ and σ was unknown.

8. $z = \dfrac{x - \mu}{\sigma} = \dfrac{20 - 25.6}{3.5} = -1.60$

$P(x < 20) = P(z < -1.60) = 0.0548$

$z = \dfrac{\bar{x} - \mu}{\dfrac{\sigma}{\sqrt{n}}} = \dfrac{20 - 25.6}{\dfrac{3.5}{\sqrt{15}}} \approx -6.20$

$P(\bar{x} < 20) = P(z < -6.20) \approx 0$

You are more likely to select one woman with a BMI less than 20.

9. (a) $\left(\dfrac{(n-1)s^2}{\chi_R^2}, \dfrac{(n-1)s^2}{\chi_L^2}\right) = \left(\dfrac{29 \cdot (3.2)^2}{45.722}, \dfrac{29 \cdot (3.2)^2}{16.047}\right) \approx (6.495, 18.506)$

(b) $\left(\sqrt{6.495}, \sqrt{18.506}\right) \approx (2.549, 4.302)$

Hypothesis Testing with One Sample

7.1 Try It Yourself Solutions

1a. (1) The mean ... is 74 months.

$\mu = 74$

(2) The variance ... is less than or equal to 3.5.

$\sigma^2 \leq 3.5$

(3) The proportion ... is greater than 39%.

$p > 0.39$

b. (1) $\mu \neq 74$ (2) $\sigma^2 > 3.5$ (3) $p \leq 0.39$

c. (1) $H_0: \mu = 74$ and $H_a: \mu \neq 74$ (Claim: H_0)

(2) $H_0: \sigma^2 \leq 3.5$ and $H_a: \sigma^2 > 3.5$ (Claim: H_0)

(3) $H_0: p \leq 0.39$ and $H_a: p > 0.39$ (Claim: H_1)

2a. $H_0: p \leq 0.01$ and $H_1: p > 0.01$

b. Type I error will occur if the actual proportion is less than or equal to 0.01, but you decided to reject H_0.

Type II error will occur if the actual proportion is greater than 0.01, but you do not reject H_0.

c. Type II error is more serious since you would be misleading the consumer possibly causing serious injury or death.

3a. (1) $H_0: \mu = 74$ and $H_a: \mu \neq 74$

(2) $H_0: p \leq 0.39$ and $H_a: p > 0.39$

b. (1) two-tailed   (2) right-tailed

c. (1) 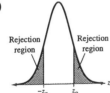 (2)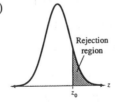

4a. There is enough evidence to reject the radio station's claim.

b. There is not enough evidence to decide that the radio station's claim is false.

5a. (No answer required)

b. (1) $\mu \leq 650$ (2) $\mu = 98.6$

6a. (1) Prove claim   (2) Disprove claim

b. (1) $H_a: \mu > 2400$ (2) $H_0: \mu \geq 2400$

7.1 EXERCISE SOLUTIONS

1. $H_0: \mu \leq 645$ and $H_a: \mu > 645$

3. $H_0: \sigma = 5$ and $H_a: \sigma \neq 5$

5. $H_0: p \geq 0.45$ and $H_a: p < 0.45$

7. c $H_a: \mu < 3$

9. b $H_a: \mu \neq 3$

11. $\mu > 750$

$H_0: \mu \leq 750$ and $H_a: \mu > 750$ (Claim: H_a)

13. $\sigma \leq 1220$

$H_0: \sigma \leq 1220$ and $H_a: \sigma > 1220$ (Claim: H_0)

15. $H_0: p = 0.44$

$H_0: p = 0.44$ and $H_a: p \neq 0.44$ (Claim: H_0)

17. Type I: Rejecting $H_0: p \geq 0.24$ when actually $p \geq 0.24$
Type II: Not rejecting $H_0: p \geq 0.24$ when actually $p < 0.24$.

19. Type I: Rejecting $H_0: \sigma \leq 23$ when actually $\sigma \leq 23$.
Type II: Not rejecting $H_0: \sigma \leq 23$ when actually $\sigma > 23$.

21. Type I: Rejecting $H_0: p \leq 0.60$ when actually $p \leq 0.60$
Type II: Not rejecting $H_0: p \leq 0.24$ when actually $p > 0.24$.

23. The null hypothesis is $H_0: p \geq 0.14$, the alternative hypothesis is $H_a: p < 0.14$. Therefore, because the alternative hypothesis contains <, the test is a left-tailed test.

25. The null hypothesis is $H_0: p = 0.90$, the alternative hypothesis is $H_a: p \neq 0.90$. Therefore, because the alternative hypothesis contains $\neq$, the test is a two-tailed test.

27. The null hypothesis is $H_0: \sigma = 0.053$, the alternative hypothesis is $H_a: \mu \neq 0.053$. Therefore, because the alternative hypothesis contains $\neq$, the test is a two-tailed test.

29. (a) There is enough evidence to reject the company's claim.

(b) There is not enough evidence to decide that the company's claim is false.

31. (a) There is enough evidence to support the Dept of Labor's claim.

(b) There is not enough evidence to decide that the Dept of Labor's claim is true.

33. (a) There is enough evidence to reject the manufacturer's claim.

(b) There is not enough evidence to reject the manufacturer's claim.

35. $\mu = 10$

37. (a) $H_0: \mu \leq 15$ and $H_a: \mu > 15$

(b) $H_0: \mu \geq 15$ and $H_a: \mu < 15$

39. If you decrease α, you are decreasing the probability that you reject H_0. Therefore, you are increasing the probability of failing to reject H_0. This could increase β, the probability of failing to reject H_0 when H_0 is false.

41. (a) Reject H_0 since the CI is located below 70.

(b) Do not reject H_0 since the CI includes values larger than 70.

(c) Do not reject H_0 since the CI includes values larger than 70.

7.2 HYPOTHESIS TESTING FOR THE MEAN $(n \geq 30)$

7.2 Try It Yourself Solutions

1a.

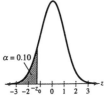

b. Area $= 0.1003$

c. $z_0 = -1.28$

2a.

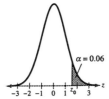

b. Area $= 0.9394$ or 0.9406

c. $z_0 = 1.55$ or 1.56

3a.

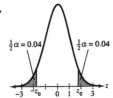

b. 0.0401 and 0.9599

c. $z_0 = -1.75$ and 1.75

4a. $H_0: \mu \geq 8.5 \ \ H_a: \mu < 8.5$ (Claim: H_a)

b. $\alpha = 0.01$

c. $z_0 = -2.33$; Rejection region: $z \leq -2.33$

d. $z = \dfrac{x - \mu}{\dfrac{s}{\sqrt{n}}} = \dfrac{8.2 - 8.5}{\dfrac{0.5}{\sqrt{35}}} \approx 3.550$

e.

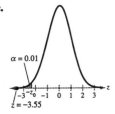

f. Reject H_0; There is enough evidence to support the claim.

5a. $\alpha = 0.01$

b. $\pm z_0 = \pm 2.575$

c. $z = -2.24$

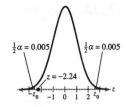

d. Fail to reject H_0; There is not enough evidence to support the claim.

6a. (1) $P = 0.0347 > 0.01 = \alpha$

(2) $P = 0.0347 < 0.05 = \alpha$

b. (1) Fail to reject H_0

(2) Reject H_0

7a. $P = 0.039 > 0.01 = \alpha$

b. Fail to reject H_0

8a. H_0: $\mu \leq 35$ and H_a: $\mu > 35$ (Claim: H_a)

b. $\alpha = 0.01$

c. $z = \dfrac{\bar{x} - \mu}{\frac{s}{\sqrt{n}}} = \dfrac{36 - 35}{\frac{4}{\sqrt{100}}} = 2.500$

d. P-value = Area right of $z = 2.50 = 0.0062$

e. Reject H_0 since P-value = $0.0062 < 0.05 = \alpha$

f. There is enough evidence to support the claim.

9a. H_0: $\mu = 150$ and H_a: $\mu \neq 150$ (Claim: H_0)

b. $\alpha = 0.01$

c. $z = \dfrac{\bar{x} - \mu}{\frac{\sigma}{\sqrt{n}}} = \dfrac{143 - 150}{\frac{15}{\sqrt{35}}} \approx -2.76$

d. P-value = Area left of $z = -2.76 = 0.0029$

e. Reject H_0 since P-value = $0.0029 < 0.01 = \alpha$

f. There is enough evidence to state the claim is false.

7.2 EXERCISE SOLUTIONS

1. Specify the level of significance, α. Decide whether the test is left-tailed, right-tailed, or two-tailed. Find the critical value(s), z_0, as follows: (a) Left-tailed: find z_0 that corresponds to an area of α. (b) Right-tailed: find z_0 that corresponds to an area of $1 - \alpha$. (c) Two-tailed: find $\pm z_0$ that corresponds to $\frac{1}{2}\alpha$ and $1 - \frac{1}{2}\alpha$.

3. 1.645

5. -1.88

7. ± 2.33

9. Right-tailed ($\alpha = 0.01$)

11. Two-tailed ($\alpha = 0.10$)

13. (a) Fail to reject H_0 since $-1.645 < z < 1.645$

(b) Reject H_0 since $z > 1.645$

(c) Fail to reject H_0 since $-1.645 < z < 1.645$

(d) Reject H_0 since $z < -1.645$

15. (a) Fail to reject H_0 since $z < 1.285$

(b) Fail to reject H_0 since $z < 1.285$

(c) Fail to reject H_0 since $z < 1.285$

(d) Reject H_0 since $z > 1.285$

17. $H_0: \mu = 40$ and $H_a: \mu \neq 40$

$\mu = 0.05 \rightarrow z_0 = \pm 1.96$

$$z = \frac{\bar{x} - \mu}{\frac{s}{\sqrt{n}}} = \frac{39.2 - 40}{\frac{3.23}{\sqrt{75}}} \approx -2.145$$

Reject H_0. There is enough evidence to reject the claim.

19. $H_0: \mu = 6000$ and $H_a: \mu \neq 6000$

$\alpha = 0.01 \rightarrow z_0 = \pm 2.575$

$$z = \frac{\bar{x} - \mu}{\frac{s}{\sqrt{n}}} = \frac{5800 - 6000}{\frac{350}{\sqrt{35}}} \approx -3.381$$

Reject H_0. There is enough evidence to support the claim.

21. (a) $H_0: \mu = 40$ and $H_a: \mu \neq 40$ (Claim: H_0)

(b) $z_0 = \pm 2.575$
Rejection regions: $z < -2.575$ and $z > 2.575$

(c) $z = \dfrac{\bar{x} - \mu}{\frac{s}{\sqrt{n}}} = \dfrac{39.2 - 40}{\frac{7.5}{\sqrt{30}}} \approx -0.584$

(d) Fail to reject H_0. There is not enough evidence to reject the claim.

23. (a) $H_0: \mu \geq 750$ and $H_a: \mu < 750$ (Claim: H_0)

(b) $z_0 = -2.05$
Rejection regions: $z < -2.05$

(c) $z = \dfrac{\bar{x} - \mu}{\frac{s}{\sqrt{n}}} = \dfrac{745 - 750}{\frac{60}{\sqrt{36}}} \approx -0.500$

(d) Fail to reject H_0. There is not enough evidence to reject the claim.

25. (a) $H_0: \mu \leq 28$ and $H_a: \mu > 28$ (Claim: H_a)

(b) $z_0 = 1.55$

Rejection regions: $z > 1.55$

(c) $\bar{x} \approx 32.861$ $s \approx 22.128$

$$z = \frac{\bar{x} - \mu}{\frac{s}{\sqrt{n}}} = \frac{32.861 - 28}{\frac{22.128}{\sqrt{36}}} \approx 1.318$$

(d) Fail to reject H_0. There is not enough evidence to support the claim.

27. (a) Fail to reject H_0.

(b) Reject H_0 $(P = 0.461 < 005 = \alpha)$

29. (a) $H_0: \mu \leq 260$ and $H_a: \mu > 260$ (Claim: H_a)

(b) $z = \dfrac{\bar{x} - \mu}{\frac{s}{\sqrt{n}}} = \dfrac{265 - 260}{\frac{55}{\sqrt{85}}} \approx 0.838$

(c) P-value = {Area to right of $z = 0.84$} = 0.2005

(d) Fail to reject H_0. There is not enough evidence to support the claim.

31. (a) $H_0: \mu \leq 7$ and $H_a: \mu > 7$ (Claim: H_a)

(b) $z = \dfrac{\bar{x} - \mu}{\frac{s}{\sqrt{n}}} = \dfrac{7.8 - 7}{\frac{2.67}{\sqrt{100}}} \approx 2.996$

(c) P-value = {Area to right of $z = 3.00$} = 0.0013

(d) Reject H_0. There is enough evidence to support the claim.

33. (a) $H_0: \mu = 15$ and $H_a: \mu \neq 15$ (Claim: H_0)

(b) $\bar{x} \approx 14.834$ $s \approx 4.288$

$$z = \frac{\bar{x} - \mu}{\frac{s}{\sqrt{n}}} = \frac{14.834 - 15}{\frac{4.288}{\sqrt{32}}} \approx -0.219$$

(c) P-value = 2{Area to left of $z = -0.22$} = 2{0.4129} = 0.8258

(d) Fail to reject H_0. There is not enough evidence to reject the claim.

35. $z = \dfrac{\bar{x} - \mu}{\frac{s}{\sqrt{n}}} = \dfrac{9900 - 10{,}000}{\frac{280}{\sqrt{30}}} \approx -1.956$

P-value = {Area left of $z = -1.96$} = 0.025

Fail to reject H_0. There is not enough evidence to support the claim.

37. Using the classical z-test, the test statistic is compared to critical values. The z-test using a P-value compares the P-value to the level of significance α.

7.3 HYPOTHESIS TESTING FOR THE MEAN $(n < 30)$

7.3 Try It Yourself Solutions

1a. 2.650

b. $t_0 = -2.650$

2a. 1.860

b. $t_0 = +1.860$

3a. 2.947

b. $t_0 = \pm 2.947$

4a. H_0: $\mu \geq \$875$ (claim) and H_a: $\mu < \$875$

b. $\alpha = 0.01$ and d.f. $= n - 1 = 8$

c. $t_0 = -2.896 \rightarrow$ Reject H_0 if $t \leq -2.896$

d. $t = \dfrac{\bar{x} - \mu}{\frac{s}{\sqrt{n}}} = \dfrac{825 - 875}{\frac{62}{\sqrt{9}}} \approx -2.419$

e. Fail to reject H_0.

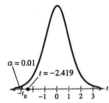

f. There is not enough evidence to reject the claim.

5a. H_0: $\mu = 1890$ (claim) and H_a: $\mu \neq 1890$

b. $\alpha = 0.01$ and d.f. $= n - 1 = 18$

c. $t_0 = \pm 2.878 \rightarrow$ Reject H_0 if $t \leq -2.878$ or $t \geq 2.878$

d. $t = \dfrac{\bar{x} - \mu}{\frac{s}{\sqrt{n}}} = \dfrac{2500 - 1890}{\frac{700}{\sqrt{19}}} \approx 3.798$

e. Reject H_0

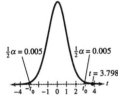

f. There is enough evidence to reject the company's claim.

6a. $t = \dfrac{\bar{x} - \mu}{\frac{s}{\sqrt{n}}} = \dfrac{126 - 134}{\frac{12}{\sqrt{6}}} \approx -1.633$

P-value $= \{$Area left of $t = -1.633\} = 0.082$

b. P-value $= 0.082 > 0.05 = \alpha$

c. Fail to reject H_0

d. There is not enough evidence to reject the claim.

7.3 EXERCISE SOLUTIONS

1. Identify the level of significance, α, and the degrees of freedom, d.f. $= n - 1$. Find the critical value(s) using the t-distribution table in the row with $n - 1$ d.f. If the hypothesis test is:

    (1) left-tailed, use "One Tail α" column with a negative sign.

    (2) right-tailed, use "One Tail α" column with a positive sign.

    (3) two-tailed, use "Two Tail α" column with a negative and a positive sign.

3. $t_0 = 1.717$

5. $t_0 = -2.101$

7. $t_0 = \pm 2.779$

9. (a) Fail to reject H_0 since $t > -2.086$.

    (b) Fail to reject H_0 since $t > -2.086$.

    (c) Fail to reject H_0 since $t > -2.086$.

    (d) Reject H_0 since $t < -2.086$.

11. (a) Fail to reject H_0 since $-2.602 < t < 2.602$.

    (b) Fail to reject H_0 since $-2.602 < t < 2.602$.

    (c) Reject H_0 since $t > 2.602$.

    (d) Reject H_0 since $t < -2.602$.

13. H_0: $\mu = 15$ (claim) and H_a: $\mu \neq 15$
    $\alpha = 0.01$ and d.f. $= n - 1 = 5$
    $t_0 = \pm 4.032$
    $$t = \frac{\bar{x} - \mu}{\frac{s}{\sqrt{n}}} = \frac{13.9 - 15}{\frac{3.23}{\sqrt{6}}} \approx -0.834$$

    Fail to reject H_0. There is not enough evidence to reject the claim.

15. H_0: $\mu \geq 8000$ (claim) and H_a: $\mu < 8000$

    $\alpha = 0.01$ and d.f. $= n - 1 = 24$

    $t_0 = -2.492$

    $$t = \frac{\bar{x} - \mu}{\frac{s}{\sqrt{n}}} = \frac{7700 - 8000}{\frac{450}{\sqrt{25}}} \approx -3.333$$

    Reject H_0. There is enough evidence to reject the claim.

17. (a) H_0: $\mu \geq 100$ and H_a: $\mu < 100$ (claim)

    (b) $t_0 = -3.747 \rightarrow$ Reject H_0 if $t < -3.747$.

    (c) $t = \frac{\bar{x} - \mu}{\frac{s}{\sqrt{n}}} = \frac{75 - 100}{\frac{12.50}{\sqrt{5}}} \approx -4.472$

    (d) Reject H_0

19. (a) $H_0: \mu \le 1$ and $H_a: \mu > 1$ (claim)

(b) $t_0 = 1.796 \rightarrow$ Reject H_0 if $t > 1.796$.

(c) $t = \dfrac{\bar{x} - \mu}{\frac{s}{\sqrt{n}}} = \dfrac{1.2 - 1}{\frac{0.3}{\sqrt{12}}} \approx 2.309$

(d) Reject H_0. There is enough evidence to support the claim.

21. (a) $H_0: \mu = \$24{,}600$ (claim) and $H_a: \mu \ne \$24{,}600$

(b) $t_0 = \pm 2.262 \rightarrow$ Reject H_0 if $t < -2.262 \ or \ t > 2.262$.

(c) $\bar{x} \approx \$24{,}124.300 \qquad s \approx \$2{,}628.935$

$t = \dfrac{\bar{x} - \mu}{\frac{s}{\sqrt{n}}} = \dfrac{24{,}124.300 - 24{,}600}{\frac{2{,}628.935}{\sqrt{10}}} \approx -0.572$

(d) Fail to reject H_0. There is not enough evidence to reject the claim.

23. (a) $H_0: \mu \ge 3.0$ and $H_a: \mu < 3.0$ (claim)

(b) $\bar{x} = 2.785 \qquad s = 0.828$

$t = \dfrac{\bar{x} - \mu}{\frac{s}{\sqrt{n}}} = \dfrac{2.785 - 3.0}{\frac{0.828}{\sqrt{20}}} \approx -1.161$

$P\text{-value} = \{\text{area left of } t = -1.161\} = 0.130$

(c) Fail to reject H_0. There is not enough evidence to reject the claim.

25. (a) $H_0: \mu \ge 32$ and $H_a: \mu < 32$ (claim)

(b) $\bar{x} = 30.167 \qquad s = 4.004$

$t = \dfrac{\bar{x} - \mu}{\frac{s}{\sqrt{n}}} = \dfrac{30.167 - 32}{\frac{4.004}{\sqrt{18}}} \approx -1.942$

$P\text{-value} = \{\text{area left of } t = -1.942\} \approx 0.034$

(c) Fail to reject H_0. There is not enough evidence to support the claim.

27. $H_0: \mu \le \$1500$ and $H_a: \mu > 1500$ (claim)

$t = \dfrac{\bar{x} - \mu}{\frac{s}{\sqrt{n}}} = \dfrac{1700 - 1500}{\frac{325}{\sqrt{6}}} \approx 1.507$

$P\text{-value} = \{\text{area right of } t = 1.507\} \approx 0.096$

Since $0.096 > 0.01 = \alpha$, fail to reject H_0.

29. Since σ is unknown, $n < 30$, and the gas mileage is normally distributed, use the t-distribution.

$H_0: \mu \ge 21$ (claim) and $H_a: \mu < 21$

$t = \dfrac{\bar{x} - \mu}{\frac{s}{\sqrt{n}}} = \dfrac{19 - 21}{\frac{4}{\sqrt{5}}} \approx 1.118$

$P\text{-value} = \{\text{area left of } t = -1.118\} = 0.163$

Fail to reject H_0. There is not enough evidence to reject the claim.

31. Since σ is known, $n < 30$, and the gas mileage is normally distributed, use the z-distribution.

$H_0: \mu \ge 21$ (claim) and $H_a: \mu < 21$

$$t = \frac{\bar{x} - \mu}{\frac{s}{\sqrt{n}}} = \frac{19 - 21}{\frac{5}{\sqrt{5}}} \approx -0.894$$

P-Value = {area left of $z = -0.89$} = 0.1867

Fail to reject H_0. There is not enough evidence to reject the claim.

33. $H_0: \mu \geq 50$ (claim) and $H_a: \mu < 50$

It is not necessary to a hypothesis test to test the repairer's claim. Note:

$$t = \frac{\bar{x} - \mu}{\frac{s}{\sqrt{n}}} = \frac{75 - 50}{\frac{12.50}{\sqrt{5}}} = 4.472$$

Recall from #17 that:

$$t = \frac{\bar{x} - \mu}{\frac{s}{\sqrt{n}}} = \frac{75 - 100}{\frac{12.50}{\sqrt{5}}} = -4.472$$

7.4 HYPOTHESIS TESTING FOR PROPORTIONS

7.4 Try It Yourself Solutions

1a. $H_0: p \geq 0.30$ and $H_a: p < 0.30$ (claim)

b. $\alpha = 0.05$

c. $z_0 = -1.645 \rightarrow$ Reject H_0 if $z < -1.645$

d. $z = \frac{\hat{p} - p}{\sqrt{\frac{pq}{n}}} = \frac{0.20 - 0.30}{\sqrt{\frac{(0.30)(0.70)}{86}}} \approx -2.024$

e. Reject H_0

f. There is enough evidence to support the claim.

2a. $H_0: p = 0.05$ (claim) and $H_a: p \neq 0.05$

b. $\alpha = 0.01$

c. $z_0 = \pm 2.575 \rightarrow$ Reject H_0 if $z < -2.575$ or $z > 2.575$

d. $z = \frac{\hat{p} - p}{\sqrt{\frac{pq}{n}}} = \frac{0.08 - 0.05}{\sqrt{\frac{(0.05)(0.95)}{250}}} \approx 2.176$

e. Fail to reject H_0

f. There is not enough evidence to reject the claim.

3a. $H_0: p \leq 0.38$ and $H_a: p > 0.38$ (claim)

b. $\alpha = 0.01$

c. $z_0 = 2.33 \rightarrow$ Reject H_0 if $z > 2.33$

d. $\hat{p} = \dfrac{x}{n} = \dfrac{33}{75} = 0.440$

$z = \dfrac{\hat{p} - p}{\sqrt{\dfrac{pq}{n}}} = \dfrac{0.440 - 0.38}{\sqrt{\dfrac{(0.38)(0.62)}{75}}} \approx 1.071$

e. Fail to reject H_0

f. There is not enough evidence to support the claim.

7.4 EXERCISE SOLUTIONS

1. Verify that $np \geq 5$ and $nq \geq 5$. State H_0 and H_a. Specify the level of significance, α. Determine the critical value(s) and rejection region(s). Find the standardized test statistic. Make a decision and interpret in the context of the original claim.

3. $np = (105)(0.25) = 26.25 \geq 5$
$nq = (105)(0.75) = 78.75 \geq 5 \rightarrow$ use normal distribution
$H_0: p = 0.25$ and $H_a: p \neq 0.25$ (claim)
$z_0 \pm 1.96$
$z = \dfrac{\hat{p} - p}{\sqrt{\dfrac{pq}{n}}} = \dfrac{0.239 - 0.25}{\sqrt{\dfrac{(0.25)(0.75)}{105}}} \approx -0.260$
Fail to reject H_0. There is not enough evidence to support the claim.

5. $np = (35)(0.60) = 21 \geq 5$
$nq = (35)(0.40) = 14 \geq 5 \rightarrow$ use normal distribution
$H_0: p \geq 0.60$ and $H_0: p < 0.60$ (claim)
$z_0 = -2.33$
$z = \dfrac{\hat{p} - p}{\sqrt{\dfrac{pq}{n}}} = \dfrac{0.58 - 0.60}{\sqrt{\dfrac{(0.60)(0.40)}{35}}} \approx -0.242$
Fail to reject H_0. There is not enough evidence to support the claim.

7. (a) $H_0: p \geq 0.25$ (claim) and $H_a: p < 0.25$

(b) $z_0 = -2.33 \rightarrow$ Reject H_0 if $z < -2.33$

(c) $z = \dfrac{\hat{p} - p}{\sqrt{\dfrac{pq}{n}}} = \dfrac{0.245 - 0.25}{\sqrt{\dfrac{(0.25)(0.75)}{200}}} \approx -0.163$

(d) Fail to reject H_0. There is not enough evidence to reject the claim.

9. (a) $H_0: p \leq 0.30$ and $H_a: p > 0.30$ (claim)

(b) $z_0 = 1.88 \rightarrow$ Reject H_0 if $z > 1.88$

(c) $z = \dfrac{\hat{p} - p}{\sqrt{\dfrac{pq}{n}}} = \dfrac{0.32 - 0.30}{\sqrt{\dfrac{(0.30)(0.70)}{1050}}} \approx 1.414$

(d) Fail to reject H_0. There is not enough evidence to support the claim.

11. (a) $H_0: p = 0.60$ (claim) and $H_a: p \neq 0.60$

(b) $z_0 = \pm 2.33 \rightarrow$ Reject H_0 if $z < -2.33$ or > 2.33

(c) $\hat{p} = \dfrac{1004}{1762} \approx 0.570$

$$z = \frac{\hat{p} - p}{\sqrt{\dfrac{pq}{n}}} = \frac{0.570 - 0.60}{\sqrt{\dfrac{(0.60)(0.40)}{1762}}} \approx -2.587$$

(d) Reject H_0. There is enough evidence to reject the claim.

13. $H_0: p \geq 0.52$ (claim) and $H_a: p < 0.52$

$z_o = -1.645$

$$z = \frac{\hat{p} - p}{\sqrt{\dfrac{pq}{n}}} = \frac{0.48 - 0.52}{\sqrt{\dfrac{(0.52)(0.48)}{30}}} \approx -0.439$$

Fail to reject H_0. There is not enough evidence to reject the claim.

15. P-values are calculated in the same manner as when using the Z-test for testing the mean.

$H_0: p \geq 0.25$ (claim) and $H_a: p < 0.25$

$z = -0.103$

P-value $= \{$area left of $z = -0.10\} = 0.4602$

Since $0.4602 > 0.01 = \alpha$, fail to reject H_0. There is not enough evidence to reject the claim.

7.5 HYPOTHESIS TESTING FOR THE VARIANCE AND STANDARD DEVIATION

7.5 Try It Yourself Solutions

1a. $\chi_0^2 = 33.409$

2a. $\chi_0^2 = 17.708$

3a. $\chi_R^2 = 31.526$

b. $\chi_L^2 = 8.231$

4a. $H_0: \sigma^2 \leq 0.40$ (claim) and $H_a: \sigma^2 > 0.40$

b. $\alpha = 0.01$ and d.f. $= n - 1 = 30$

c. $\chi_0^2 = 50.892 \rightarrow$ Reject H_0 if $\chi^2 > 50.892$

d. $\chi^2 = \dfrac{(n-1)s^2}{\sigma^2} = \dfrac{(30)(0.75)}{0.40} = 56.250$

e. Reject H_0

f. There is enough evidence to reject the claim.

5a. $H_0: \sigma \geq 3.7$ and $H_a: \sigma < 3.7$ (claim)

b. $\alpha = 0.05$ and d.f. $= n - 1 = 8$

c. $\chi_0^2 = 2.733 \rightarrow$ Reject H_0 if $\chi^2 < 2.733$

d. $\chi^2 = \dfrac{(n-1)s^2}{\sigma^2} = \dfrac{(8)(3.0)^2}{(3.7)^2} \approx 5.259$

e. Fail to reject H_0

f. There is not enough evidence to support the claim.

6a. H_0: $\sigma^2 = 8.6$ (claim) and H_a: $\sigma^2 \neq 8.6$

b. $\alpha = 0.01$ and d.f. $= n - 1 = 9$

c. $\chi_L^2 = 1.735$ and $\chi_R^2 \approx 23.589$
   Reject H_0 if $\chi^2 > 23.589$ or $\chi^2 < 1.735$

d. $\chi^2 = \dfrac{(n-1)s^2}{\sigma^2} = \dfrac{(9)(4.3)}{(8.6)} = 4.500$

e. Fail to reject H_0

f. There is not enough evidence to reject the claim.

7.5 EXERCISE SOLUTIONS

1. Specify the level of significance, α. Determine the degrees of freedom. Determine the critical values using the χ^2 distribution. If (a) right-tailed test, use the value that corresponds to d.f and α. (b) left-tailed test, use the value that corresponds to d.f and $1 - \alpha$. (c) two-tailed test, use the value that corresponds to d.f and $\frac{1}{2}\alpha$ and $1 - \frac{1}{2}\alpha$.

3. $\chi_0^2 = 38.885$

5. $\chi_0^2 = 0.872$

7. $\chi_L^2 = 7.261$ $\chi_R^2 = 24.996$

9. (a) Fail to reject H_0 (b) Fail to reject H_0

   (c) Fail to reject H_0 (d) Reject H_0

11. (a) Fail to reject H_0 (b) Reject H_0

   (c) Reject H_0 (d) Fail to reject H_0

13. H_0: $\sigma^2 = 0.52$ (claim); H_a: $\sigma^2 \neq 0.52$
    $\chi_L^2 = 7.564$ $\chi_R^2 = 30.191$
    $\chi^2 = \dfrac{(n-1)s^2}{\sigma^2} = \dfrac{(17)(0.508)}{0.52} \approx 16.608$
    Fail to reject H_0. There is not enough evidence to reject the claim.

15. H_0: $\sigma \geq 40$ and H_a: $\sigma < 40$ (claim)
    $\chi_0^2 = 3.053$
    $\chi^2 = \dfrac{(n-1)s^2}{\sigma^2} = \dfrac{(11)(40.8)^2}{(40)^2} \approx 11.444$
    Fail to reject H_0. There is not enough evidence to support the claim.

17. (a) H_0: $\sigma^2 = 3$ (claim) and H_a: $\sigma^2 \neq 3$

   (b) $\chi_L^2 = 13.844$ $\chi_R^2 = 41.923$
       Reject H_0 if $\chi^2 > 41.923$ or $\chi^2 < 13.844$

   (c) $\chi^2 = \dfrac{(n-1)s^2}{\sigma^2} = \dfrac{(26)(2.8)}{3} \approx 24.267$

   (d) Fail to reject H_0. There is not enough evidence to support the claim.

19. (a) H_0: $\sigma \geq 29$ and H_a: $\sigma < 29$ (claim)

(b) $\chi_0^2 = 13.240 \rightarrow$ Reject H_0 if $\chi^2 < 13.240$

(c) $\chi^2 = \dfrac{(n-1)s^2}{\sigma^2} = \dfrac{(21)(27.7)^2}{(29)^2} \approx 19.159$

(d) Fail to reject H_0. There is not enough evidence to support the claim.

21. (a) H_0: $\sigma \leq 0.5$ (claim) and H_a: $\sigma > 0.5$

(b) $\chi_0^2 = 33.196 \rightarrow$ Reject H_0 if $\chi^2 > 33.196$

(c) $\chi^2 = \dfrac{(n-1)s^2}{\sigma^2} = \dfrac{(24)(0.7)^2}{(0.5)^2} = 47.040$

(d) Reject H_0. There is enough evidence to reject the claim.

23. (a) H_0: $\sigma^2 \leq 20{,}000$ and H_a: $\sigma^2 > 20{,}000$ (claim)

(b) $\chi_0^2 = 24.996 \rightarrow$ Reject H_0 if $\chi^2 > 24.996$

(c) $s = 20{,}662.992$

$$\chi^2 = \dfrac{(n-1)s^2}{\sigma^2} = \dfrac{(15)(20{,}662.992)^2}{(20{,}000)^2} \approx 16.011$$

(d) Fail to reject H_0. There is not enough evidence to support the claim.

25. $\chi^2 = 16.011$

P-value = {area right of $\chi^2 = 16.011$} = 0.381

Fail to reject H_0.

CHAPTER 7 REVIEW EXERCISE SOLUTIONS

1. H_0: $\mu \leq 1593$ (claim) and H_a: $\mu > 1593$

3. H_0: $\mu = 150{,}020$ and H_a: $\mu \neq 150{,}020$ (claim)

5. (a) H_0: $p = 0.85$ (claim) and H_a: $p \neq 0.85$

(b) Type I error will occur if H_0 is rejected when the actual proportion of American adults who use nonprescription pain relievers is 0.85.

Type II error if H_0 is not rejected when the actual proportion of American adults who use nonprescription pain relievers is not 0.85.

(c) Two-tailed, since hypothesis compares "= vs ≠"

(d) There is enough evidence to reject the claim.

(e) There is not enough evidence to reject the claim.

7. (a) H_0: $\mu \leq 50$ (claim) and H_a: $\mu > 50$

(b) Type I error will occur if H_0 is rejected when the actual standard deviation sodium content is no more than 50 mg.

Type II error if H_0 is not rejected when the actual standard deviation sodium content is more than 50 mg.

(c) Right-tailed, since hypothesis compares "≤ vs >"

(d) There is enough evidence to reject the claim.

(e) There is not enough evidence to reject the claim.

9. $z_0 \approx -2.05$

11. $z_0 = 1.96$

13. $H_0: \mu \leq 45$ (claim) and $H_a: \mu > 45$

$z_0 = 1.645$

$$z = \frac{\bar{x} - \mu}{\frac{s}{\sqrt{n}}} = \frac{47.2 - 45}{\frac{6.7}{\sqrt{42}}} \approx 2.128$$

Reject H_0. There is enough evidence to reject the claim.

15. $H_0: \mu \geq 5.500$ and $H_a: \mu < 5.500$ (claim)

$z_0 = -2.33$

$$z = \frac{\bar{x} - \mu}{\frac{s}{\sqrt{n}}} = \frac{5.497 - 5.500}{\frac{0.011}{\sqrt{36}}} \approx -1.636$$

Fail to reject H_0. There is not enough evidence to support the claim.

17. $H_0: \mu \leq 0.05$ (claim) and $H_a: \mu > 0.05$

$$z = \frac{\bar{x} - \mu}{\frac{s}{\sqrt{n}}} = \frac{0.057 - 0.05}{\frac{0.018}{\sqrt{32}}} \approx 2.200$$

P-value = {are right of $z = 2.20$} = 0.0139

$\alpha = 0.10 \rightarrow$ Reject H_0

$\alpha = 0.05 \rightarrow$ Reject H_0

$\alpha = 0.01 \rightarrow$ Fail to reject H_0

19. $t_0 = \pm 2.093$

21. $t_0 = -1.345$

23. $H_0: \mu = 95$ and $H_a: \mu \neq 95$ (claim)

$t_0 = \pm 2.201$

$$t = \frac{\bar{x} - \mu}{\frac{s}{\sqrt{n}}} = \frac{94.1 - 95}{\frac{1.53}{\sqrt{12}}} \approx -2.038$$

Fail to reject H_0 There is not enough evidence to support the claim.

25. $H_0: \mu \geq 0$ (claim) and $H_a: \mu < 0$

$t_0 = -1.341$

$$t = \frac{\bar{x} - \mu}{\frac{s}{\sqrt{n}}} = \frac{-0.45 - 0}{\frac{1.38}{\sqrt{16}}} \approx -1.304$$

Fail to reject H_0. There is not enough evidence to reject the claim.

27. $H_0: \mu = \$25$ (claim) and $H_a: \mu \neq \$25$

$t_0 = \pm 1.740$

$$t = \frac{\bar{x} - \mu}{\frac{s}{\sqrt{n}}} = \frac{26.25 - 25}{\frac{3.23}{\sqrt{18}}} \approx 1.642$$

Fail to reject H_0. There is not enough evidence to reject the claim.

29. H_0: $\mu \geq 4$ (claim) and H_a: $\mu < 4$

$t_0 = -2.539$

$\bar{x} = 3.885$, $s = 1.008$

$$t = \frac{\bar{x} - \mu}{\frac{s}{\sqrt{n}}} = \frac{3.885 - 4}{\frac{1.008}{\sqrt{20}}} \approx -0.510$$

Fail to reject H_0. There is not enough evidence to reject the claim.

31. H_0: $p = 0.15$ (claim) and H_a: $p \neq 0.15$

$z_0 = \pm 1.96$

$$z = \frac{\hat{p} - p}{\sqrt{\frac{pq}{n}}} = \frac{0.09 - 0.15}{\sqrt{\frac{(0.15)(0.85)}{40}}} \approx -1.063$$

Fail to reject H_0. There is not enough evidence to reject the claim.

33. Because $np = 3.6$ is less than 5, the normal distribution cannot be used to approximate the binomial distribution.

35. H_0: $p \leq 0.40$ and H_a: $p > 0.40$ (claim)

$z_0 = 1.282$

$$\hat{p} = \frac{x}{n} = \frac{456}{1036} \approx 0.440$$

$$z = \frac{\hat{p} - p}{\sqrt{\frac{pq}{n}}} = \frac{0.440 - 0.40}{\sqrt{\frac{(0.40)(0.60)}{1036}}} \approx 2.628$$

Reject H_0. There is enough evidence to support the claim.

37. $\chi_R^2 = 30.144$

39. $\chi_R^2 = 33.196$

41. H_0: $\sigma^2 \leq 2$ and H_a: $\sigma^2 > 2$ (claim)

$\chi_0^2 = 24.769$

$$\chi^2 = \frac{(n-1)s^2}{\sigma^2} = \frac{(17)(2.38)}{(2)} = 20.230$$

Fail to reject H_0. There is not enough evidence to support the claim.

43. H_0: $\sigma^2 = 1.25$ (claim) and H_a: $\sigma \neq 1.25$

$\chi_L^2 = 0.831$ $\chi_R^2 = 12.833$

$$\chi^2 = \frac{(n-1)s^2}{\sigma^2} = \frac{(5)(1.03)^2}{(1.25)^2} \approx 3.395$$

Fail to reject H_0. There is not enough evidence to reject the claim.

45. H_0: $\sigma^2 \leq 0.01$ (claim) and H_a: $\sigma^2 > 0.01$

$\chi_0^2 = 49.645$

$$\chi^2 = \frac{(n-1)s^2}{\sigma^2} = \frac{(27)(0.064)}{(0.01)} = 172.800$$

Reject H_0. There is not enough evidence to reject the claim.

CHAPTER 7 QUIZ SOLUTIONS

1. (a) $H_0: \mu \geq 94$ (claim) and $H_a: \mu < 94$

   (b) Type I error occurs if the H_0 is rejected when actually the mean consumption is at least 94 pounds.

   Type II error occurs if the H_0 has not been rejected when actually the mean consumption is less than 94 pounds.

   (c) "$\geq$ vs $<$" $\rightarrow$ Left tailed

   σ^2 is unknown and $n \geq 30 \rightarrow z$-test

   (d) $z_0 = -2.05 \rightarrow$ Reject H_0 if $z < -2.05$

   (e) $z = \dfrac{\bar{x} - \mu}{\frac{s}{\sqrt{n}}} = \dfrac{93.5 - 94}{\frac{30}{\sqrt{103}}} \approx -0.169$

   (f) Fail to reject H_0. There is not enough evidence to reject the claim.

2. (a) $H_0: \mu \geq 25$ (claim) and $H_a: \mu < 25$

   (b) Type I error occurs if the H_0 is rejected when actually the mean mpg is at least 25.

   Type II error occurs if the H_0 has not been rejected when actually the mean mpg is less than 25.

   (c) "$\geq$ vs $<$" $\rightarrow$ Left tailed

   σ^2 is unknown and $n < 30 \rightarrow t$-test

   (d) $t_0 = -1.895 \rightarrow$ Reject H_0 if $t < -1.895$

   (e) $= \dfrac{\bar{x} - \mu}{\frac{s}{\sqrt{n}}} = \dfrac{23 - 25}{\frac{5}{\sqrt{8}}} \approx -1.131$

   (f) Fail to reject H_0. There is not enough evidence to reject the claim.

3. (a) $H_0: p \leq 0.10$ (claim) and $H_a: p > 0.10$

   (b) Type I error occurs if the H_0 is rejected when actually the proportion of microwaves needing repair is no more than 0.10.

   Type II error occurs if the H_0 has not been rejected when actually the proportion of microwaves needing repair is more than 0.10.

   (c) "$\leq$ vs $>$" $\rightarrow$ Right tailed

   $np \geq 5$ and $nq \geq 5 \rightarrow z$-test

   (d) $z_0 = 1.75$

   (e) $z = \dfrac{\hat{p} - p}{\sqrt{\frac{pq}{n}}} = \dfrac{0.13 - 0.10}{\sqrt{\frac{(0.10)(0.90)}{57}}} \approx 0.755$

   (f) Fail to reject H_0. There is not enough evidence to reject the claim.

4. (a) $H_0: \sigma = 105$ (claim) and $H_a: \sigma \neq 105$

   (b) Type I error occurs if the H_0 is rejected when actually the standard deviation of the scores is 105.

   Type II error occurs if the H_0 has not been rejected when actually the standard deviation of the scores is not 105.

(c) "$= vs \neq$" → Two-tailed

Assuming the scores are normally distributed and you are testing the hypothesized standard deviation → χ^2 test

(d) $\chi_L^2 = 3.565$ $\chi_R^2 = 29.819$

(e) $\chi^2 = \dfrac{(n-1)s^2}{\sigma^2} = \dfrac{(13)(113)^2}{(105)^2} \approx 15.056$

(f) Fail to reject H_0. There is not enough evidence to reject the claim.

5. (a) $H_0: \mu = \$53,102$ (claim) and $H_a: \mu \neq \$53,102$

(b) Type I error occurs if the H_0 is rejected when actually the mean salary is $53,102.

Type II error occurs if the H_0 has not been rejected when actually the mean salary is not $53,102.

(c) "$= vs \neq$" → Two-tailed

σ is unknown, $n < 30$, and assuming the salaries are normally distributed → t-test

(d) Reject H_0 if P-value $\leq \alpha = 0.05$.

(e) $t = \dfrac{\bar{x} - \mu}{\dfrac{s}{\sqrt{n}}} = \dfrac{52,201 - 53,102}{\dfrac{6500}{\sqrt{12}}} \approx -0.480$

P-value $= 2\{\text{area left of } t = -0.480\} = 2(0.320) = 0.640$

(f) Fail to reject H_0. There is not enough evidence to reject the claim.

Hypothesis Testing with Two Samples

8.1 TESTING THE DIFFERENCE BETWEEN TWO MEANS

8.1 Try It Yourself Solutions

1a. $H_0: \mu_1 = \mu_2$ and $H_1: \mu_1 \neq \mu_2$ (claim)

b. $\alpha = 0.01$

c. $z_0 = \pm 2.575 \rightarrow$ Reject H_0 if $z > 2.575$ or $z < -2.575$

d. $z = \dfrac{(\bar{x}_1 - \bar{x}_2) - (\mu_1 - \mu_2)}{\sqrt{\dfrac{s_1^2}{n_1} + \dfrac{s_2^2}{n_2}}} = \dfrac{(3900 - 3500) - (0)}{\sqrt{\dfrac{(900)^2}{50} + \dfrac{(500)^2}{50}}} \approx 2.747$

e. Reject H_0

f. There is enough evidence to support the claim.

2a. $z = \dfrac{(\bar{x}_1 - \bar{x}_2) - (\mu_1 - \mu_2)}{\sqrt{\dfrac{s_1^2}{n_1} + \dfrac{s_2^2}{n_2}}} = \dfrac{(252 - 244) - (0)}{\sqrt{\dfrac{(22)^2}{150} + \dfrac{(18)^2}{200}}} \approx 3.634$

$\rightarrow P$-value = {area right of $z = 3.634$} = 0.000140

b. Rejection region is $z > 1.645$ or P-value $< 0.05 = \alpha$

c. Reject H_0

8.1 EXERCISE SOLUTIONS

1. State the hypotheses and identify the claim. Specify the level of significance and find the critical value(s). Find the standardized test statistic. Make a decision and interpret in the context of the claim.

3. $H_0: \mu_1 = \mu_2$ (claim) and $H_1: \mu_1 \neq \mu_2$

Rejection regions: $z_0 < -1.96$ and $z_0 > 1.96$ (Two-tailed test)

(a) $\bar{x}_1 - \bar{x}_2 = 16 - 14 = 2$

(b) $z = \dfrac{(\bar{x}_1 - \bar{x}_2) - (\mu_1 - \mu_2)}{\sqrt{\dfrac{s_1^2}{n_1} + \dfrac{s_2^2}{n_2}}} = \dfrac{(16 - 14) - (0)}{\sqrt{\dfrac{(1.1)^2}{50} + \dfrac{(1.5)^2}{50}}} \approx 7.603$

(c) z is in the rejection region because $7.603 > 1.96$.

(d) Reject H_0. There is enough evidence to reject the claim.

5. $H_0: \mu_1 \geq \mu_2$ and $H_1: \mu_1 < \mu_2$ (claim)

Rejection region: $z_0 < -2.33$ (Left-tailed test)

(a) $\bar{x}_1 - \bar{x}_2 = 1225 - 1195 = 30$

(b) $z = \dfrac{(\bar{x}_1 - \bar{x}_2) - (\mu_1 - \mu_2)}{\sqrt{\dfrac{s_1^2}{n_1} + \dfrac{s_2^2}{n_2}}} = \dfrac{(1225 - 1195) - (0)}{\sqrt{\dfrac{(75)^2}{35} + \dfrac{(105)^2}{105}}} \approx 1.84$

(c) z is not in the rejection region because $1.84 > -2.330$

(d) Fail to reject H_0. There is not enough evidence to support the claim.

7. (a) H_0: $\mu_1 = \mu_2$ (claim) and H_1: $\mu_1 \neq \mu_2$

(b) $z_0 = \pm 1.645 \rightarrow$ *Reject H_0 if $z < -1.645$ or $z > 1.645$*

(c) $z = \dfrac{(\bar{x}_1 - \bar{x}_2) - (\mu_1 - \mu_2)}{\sqrt{\dfrac{s_1^2}{n_1} + \dfrac{s_2^2}{n_2}}} = \dfrac{(42 - 45) - (0)}{\sqrt{\dfrac{(4.7)^2}{35} + \dfrac{(4.3)^2}{35}}} \approx -2.786$

(d) Reject H_0. There is enough evidence to reject the claim.

9. (a) H_0: $\mu_1 \geq \mu_2$ and H_1: $\mu_1 < \mu_2$ (claim)

(b) $z_0 = -2.33 \rightarrow$ Reject H_0 if $z < -2.33$

(c) $z = \dfrac{(\bar{x}_1 - \bar{x}_2) - (\mu_1 - \mu_2)}{\sqrt{\dfrac{s_1^2}{n_1} + \dfrac{s_2^2}{n_2}}} = \dfrac{(75 - 80) - (0)}{\sqrt{\dfrac{(12.50)^2}{47} + \dfrac{(20)^2}{55}}} \approx -1.536$

(d) Fail to reject H_0. There is not enough evidence to support the claim. Do not buy Model A.

11. (a) H_0: $\mu_1 = \mu_2$ (claim) and H_1: $\mu_1 \neq \mu_2$

(b) $z_0 = \pm 2.575 \rightarrow$ Reject H_0 if $z < -2.575$ or $z > 2.575$

(c) $z = \dfrac{(\bar{x}_1 - \bar{x}_2) - (\mu_1 - \mu_2)}{\sqrt{\dfrac{s_1^2}{n_1} + \dfrac{s_2^2}{n_2}}} = \dfrac{(21.2 - 20.9) - (0)}{\sqrt{\dfrac{(4.9)^2}{43} + \dfrac{(4.6)^2}{56}}} \approx 0.310$

(d) Fail to reject H_0. There is not enough evidence to reject the claim.

13. (a) H_0: $\mu_1 \leq \mu_2$ and H_1: $\mu_1 > \mu_2$ (claim)

(b) $z_0 = 1.96 \rightarrow$ Reject H_0 if $z > 1.96$

(c) $\bar{x}_1 \approx 2.130, s_1 \approx 0.490, n_1 = 30$
$\bar{x}_2 \approx 1.593, s_2 \approx 0.328, n_2 = 30$

$z = \dfrac{(\bar{x}_1 - \bar{x}_2) - (\mu_1 - \mu_2)}{\sqrt{\dfrac{s_1^2}{n_1} + \dfrac{s_2^2}{n_2}}} = \dfrac{(2.130 - 1.593) - (0)}{\sqrt{\dfrac{(0.490)^2}{30} + \dfrac{(0.328)^2}{30}}} \approx 4.988$

(d) Reject H_0. There is enough evidence to support the claim.

15. (a) H_0: $\mu_1 = \mu_2$ (claim) and H_1: $\mu_1 \neq \mu_2$

(b) $z_0 = \pm 2.575 \rightarrow$ Reject H_0 if $z < -2.575$ or $z > 2.575$

(c) $\bar{x}_1 \approx 0.875, s_1 \approx 0.011, n_1 = 35$
$\bar{x}_2 \approx 0.701, s_2 \approx 0.011, n_2 = 35$

$z = \dfrac{(\bar{x}_1 - \bar{x}_2) - (\mu_1 - \mu_2)}{\sqrt{\dfrac{s_1^2}{n_1} + \dfrac{s_2^2}{n_2}}} = \dfrac{(0.875 - 0.701) - (0)}{\sqrt{\dfrac{(0.011)^2}{35} + \dfrac{(0.011)^2}{35}}} \approx 66.172$

(d) Reject H_0. There is not enough evidence to support the claim.

17. They are equivalent through algebraic manipulation of the equation.

$$\mu_1 = \mu_2 \rightarrow \mu_1 - \mu_2 = 0$$

19. $H_0: \mu_1 - \mu_2 = -9$ (claim) and $H_1: \mu_1 - \mu_2 \neq -9$

$z_0 = \pm 2.575 \rightarrow$ Reject H_0 if $z < -2.575$ or $z > 2.575$

$$z = \frac{(\bar{x}_1 - \bar{x}_2) - (\mu_1 - \mu_2)}{\sqrt{\frac{s_1^2}{n_1} + \frac{s_2^2}{n_2}}} = \frac{(11.5 - 20) - (-9)}{\sqrt{\frac{(3.8)^2}{70} + \frac{(6.7)^2}{65}}} \approx 0.528$$

Fail to reject H_0. There is not enough evidence to reject the claim.

21. $H_0: \mu_1 - \mu_2 \leq 6000$ and $H_1: \mu_1 - \mu_2 > 6000$ (claim)

$z_0 = 1.28 \rightarrow$ Reject H_0 if $z > 1.28$

$$z = \frac{(\bar{x}_1 - \bar{x}_2) - (\mu_1 - \mu_2)}{\sqrt{\frac{s_1^2}{n_1} + \frac{s_2^2}{n_2}}} = \frac{(43{,}300 - 37{,}400) - (6000)}{\sqrt{\frac{(7800)^2}{45} + \frac{(7400)^2}{37}}} \approx -0.059$$

Fail to reject H_0. There is not enough evidence to support the claim.

23. $(\bar{x}_1 - \bar{x}_2) - z_c \sqrt{\frac{s_1^2}{n_1} + \frac{s_2^2}{n_2}} < \mu_1 - \mu_2 < (\bar{x}_1 - \bar{x}_2) + z_c \sqrt{\frac{s_1^2}{n_1} + \frac{s_2^2}{n_2}}$

$(3.2 - 4.1) - 1.96 \sqrt{\frac{(3.3)^2}{42} + \frac{(3.9)^2}{42}} < \mu_1 - \mu_2 < (3.2 - 4.1) + 1.96 \sqrt{\frac{(3.3)^2}{42} + \frac{(3.9)^2}{42}}$

$-2.45 < \mu_1 - \mu_2 < 0.65$

25. $H_0: \mu_1 - \mu_2 \leq 0$ and $H_1: \mu_1 - \mu_2 > 0$ (claim)

$z_0 = 1.645 \rightarrow$ Reject H_0 if $z > 1.645$

$$z = \frac{(\bar{x}_1 - \bar{x}_2) - (\mu_1 - \mu_2)}{\sqrt{\frac{s_1^2}{n_1} + \frac{s_2^2}{n_2}}} = \frac{(3.2 - 4.1) - (0)}{\sqrt{\frac{(3.3)^2}{42} + \frac{(3.9)^2}{42}}} = -1.14$$

Fail to reject H_0. There is not enough evidence to support the claim.

27. $H_0: \mu_1 - \mu_2 \leq 0$ and $H_1: \mu_1 - \mu_2 > 0$ (claim)

Since 0 is contained in the 95% CI for $\mu_1 - \mu_2$, fail to reject H_0. There is not enough evidence to support the claim.

8.2 TESTING THE DIFFERENCE BETWEEN TWO MEANS (SMALL INDEPENDENT SAMPLES)

8.2 Try It Yourself Solutions

1a. $H_0: \mu_1 = \mu_2$ and $H_a: \mu_1 \neq \mu_2$ (claim)

b. $\alpha = 0.05$

c. d.f. $= \min\{n_1 - 1, n_2 - 1\} = \min\{10 - 1, 12 - 1\} = 9$

d. $t_0 = \pm 2.262 \rightarrow$ Reject H_0 if $t < -2.262$ or $t > 2.262$

e. $z = \frac{(\bar{x}_1 - \bar{x}_2) - (\mu_1 - \mu_2)}{\sqrt{\frac{s_1^2}{n_1} + \frac{s_2^2}{n_2}}} = \frac{(102 - 94) - (0)}{\sqrt{\frac{(10)^2}{10} + \frac{(4)^2}{12}}} \approx 2.376$

f. Reject H_0.

g. There is enough evidence to support the claim.

2a. $H_0: \mu_1 \geq \mu_2$ and $H_a: \mu_1 < \mu_2$ (claim)

b. $\alpha = 0.10$

c. d.f. $= n_1 = n_2 - 2 = 12 + 15 - 2 = 25$

d. $t_0 = -1.316 \rightarrow$ Reject H_0 if $t < -1.316$

e. $t = \dfrac{(\bar{x}_1 - \bar{x}_2) - (\mu_1 - \mu_2)}{\sqrt{\dfrac{(n_1 - 1)s_1^2 + (n_2 - 1)s_2^2}{n_1 + n_2 - 2}} \sqrt{\dfrac{1}{n_1} + \dfrac{1}{n_2}}} = \dfrac{(73 - 74) - (0)}{\sqrt{\dfrac{(12 - 1)(2.4)^2 + (15 - 1)(3.2)^2}{12 + 15 - 2}} \sqrt{\dfrac{1}{12} + \dfrac{1}{15}}} \approx -0.90$

f. Fail to reject H_0

g. There is not enough evidence to support the claim.

8.2 EXERCISE SOLUTIONS

1. State hypotheses and identify the claim. Specify the level of significance. Determine the degrees of freedom. Find the critical value(s) and identify the rejection region(s). Find the standardized test statistic. Make a decision and interpret in the context of the original claim.

3. (a) d.f. $= n_1 + n_2 - 2 = 20$

   $t_0 = \pm 1.725$

   (b) d.f. $= \min\{n_1 - 1, n_2 - 1\} = 9$

   $t_0 = \pm 1.833$

5. (a) d.f. $= n_1 + n_2 - 2 = 22$

   $t_0 = -2.074$

   (b) d.f. $= \min\{n_1 - 1, n_2 - 1\} = 8$

   $t_0 = -2.306$

7. $H_0: \mu_1 = \mu_2$ (claim) and $H_a: \mu_1 \neq \mu_2$

   d.f. $= n_1 + n_2 - 2 = 15$

   $t_0 = \pm 2.947$ (Two-tailed test)

   (a) $\bar{x}_1 - \bar{x}_2 = 33.7 - 35.5 = -1.8$

   (b) $t = \dfrac{(\bar{x}_1 - \bar{x}_2) - (\mu_1 - \mu_2)}{\sqrt{\dfrac{(n_1 - 1)s_1^2 + (n_2 - 1)s_2^2}{n_1 + n_2 - 2}} \sqrt{\dfrac{1}{n_1} + \dfrac{1}{n_2}}} = \dfrac{(33.7 - 35.5) - (0)}{\sqrt{\dfrac{(10 - 1)(3.5)^2 + (7 - 1)(2.2)^2}{10 + 7 - 2}} \sqrt{\dfrac{1}{10} + \dfrac{1}{7}}} \approx -1.199$

   (c) t is not in the rejection region.

   (d) Fail to reject H_0. There is not enough evidence to reject the claim.

9. $H_0: \mu_1 \leq \mu_2$ (claim) and $H_a: \mu_1 > \mu_2$

   d.f. $= \min\{n_1 - 1, n_2 - 1\} = 4$

   $t_0 = 2.132$ (Right-tailed test)

   (a) $\bar{x}_1 - \bar{x}_2 = 2250 - 2305 = -55$

   (b) $t = \dfrac{(\bar{x}_1 - \bar{x}_2) - (\mu_1 - \mu_2)}{\sqrt{\dfrac{s_1^2}{n_1} + \dfrac{s_2^2}{n_2}}} = \dfrac{(2250 - 2305) - (0)}{\sqrt{\dfrac{(175)^2}{13} + \dfrac{(52)^2}{10}}} \approx -1.073$

(c) t is not in the rejection region.

(d) Fail to reject H_0. There is not enough evidence to reject the claim.

11. (a) $H_0: \mu_1 = \mu_2$ (claim) and $H_a: \mu_1 \neq \mu_2$

(b) d.f. $= n_1 + n_2 - 2 = 35$

$t_0 = \pm 1.645 \rightarrow$ Reject H_0 if $t < -1.645$ or $t > 1.645$

(c) $t = \dfrac{(\bar{x}_1 - \bar{x}_2) - (\mu_1 - \mu_2)}{\sqrt{\dfrac{(n_1 - 1)s_1^2 + (n_2 - 1)s_2^2}{n_1 + n_2 - 2}}\sqrt{\dfrac{1}{n_1} + \dfrac{1}{n_2}}}$

$= \dfrac{(23.1 - 25.3) - (0)}{\sqrt{\dfrac{(14 - 1)(8.69)^2 + (23 - 1)(7.21)^2}{14 + 23 - 2}}\sqrt{\dfrac{1}{14} + \dfrac{1}{23}}} \approx -0.833$

(d) Fail to reject H_0. There is not enough evidence to reject the claim.

13. (a) $H_0: \mu_1 \geq \mu_2$ and $H_a: \mu_1 < \mu_2$ (claim)

(b) d.f. $= \min\{n_1 - 1, n_2 - 1\} = 13$

$t_0 = -1.350 \rightarrow$ Reject H_0 if $t < -1.350$

(c) $z = \dfrac{(\bar{x}_1 - \bar{x}_2) - (\mu_1 - \mu_2)}{\sqrt{\dfrac{s_1^2}{n_1} + \dfrac{s_2^2}{n_2}}} = \dfrac{(574 - 734) - (0)}{\sqrt{\dfrac{(185)^2}{14} + \dfrac{(268)^2}{23}}} \approx -2.144$

(d) Reject H_0. There is enough evidence to support the claim.

15. (a) $H_0: \mu_1 \leq \mu_2$ and $H_a: \mu_1 > \mu_2$ (claim)

(b) d.f. $= \min\{n_1 - 1, n_2 - 1\} = 14$

$t_0 = 1.761$

(c) $z = \dfrac{(\bar{x}_1 - \bar{x}_2) - (\mu_1 - \mu_2)}{\sqrt{\dfrac{s_1^2}{n_1} + \dfrac{s_2^2}{n_2}}} = \dfrac{(30{,}800 - 25{,}700) - (0)}{\sqrt{\dfrac{(8600)^2}{19} + \dfrac{(5500)^2}{15}}} \approx 2.098$

(d) Reject H_0. There is enough evidence to support the claim.

17. (a) $H_0: \mu_1 = \mu_2$ and $H_a: \mu_1 \neq \mu_2$ (claim)

(b) d.f. $= n_1 + n_2 - 2 = 21$

$t_0 = \pm 2.831$

(c) $\bar{x}_1 = 340.300, s_1 = 22.301, n_1 = 10$

$\bar{x}_2 = 389.538, s_2 = 14.512, n_2 = 13$

$t = \dfrac{(\bar{x}_1 - \bar{x}_2) - (\mu_1 - \mu_2)}{\sqrt{\dfrac{(n_1 - 1)s_1^2 + (n_2 - 1)s_2^2}{n_1 + n_2 - 2}}\sqrt{\dfrac{1}{n_1} + \dfrac{1}{n_2}}}$

$= \dfrac{(340.300 - 389.538) - (0)}{\sqrt{\dfrac{(10 - 1)(22.301)^2 + (13 - 1)(14.512)^2}{10 + 13 - 2}}\sqrt{\dfrac{1}{10} + \dfrac{1}{13}}} \approx -6.410$

(d) Reject H_0. There is enough evidence to support the claim.

19. (a) $H_0: \mu_1 \geq \mu_2$ and $H_a: \mu_1 < \mu_2$ (claim)

(b) d.f. $= n_1 + n_2 - 2 = 42$

$t_0 = -1.282$

(c) $\bar{x}_1 = 56.684$, $s_1 = 6.961$, $n_1 = 19$

$\bar{x}_2 = 67.400$, $s_2 = 9.014$, $n_2 = 25$

$$t = \frac{(\bar{x}_1 - \bar{x}_2) - (\mu_1 - \mu_2)}{\sqrt{\dfrac{(n_1 - 1)s_1^2 + (n_2 - 1)s_2^2}{n_1 + n_2 - 2}} \sqrt{\dfrac{1}{n_1} + \dfrac{1}{n_2}}}$$

$$= \frac{(56.684 - 67.400) - (0)}{\sqrt{\dfrac{(19 - 1)(6.961)^2 + (25 - 1)(9.014)^2}{19 + 25 - 2}} \sqrt{\dfrac{1}{19} + \dfrac{1}{25}}} \approx -4.295$$

(d) Reject H_0. There is enough evidence to support the claim and to recommend changing to the new method.

21. $\hat{\sigma} = \sqrt{\dfrac{(n_1 - 1)s_1^2 + (n_2 - 1)s_2^2}{n_1 + n_2 - 2}} = \sqrt{\dfrac{(15 - 1)(6.2)^2 + (12 - 1)(8.1)^2}{15 + 12 - 2}} \approx 7.099$

$(\bar{x}_1 - \bar{x}_2) \pm t_c\hat{\sigma}\sqrt{\dfrac{1}{n_1} + \dfrac{1}{n_2}} \rightarrow (230 - 240) \pm 2.060 \cdot 7.099 \sqrt{\dfrac{1}{15} + \dfrac{1}{12}} \rightarrow$

$-10 \pm 5.664 \rightarrow -15.664 < \mu_1 - \mu_2 < -4.336$

23. $(\bar{x}_1 - \bar{x}_2) \pm t_c\sqrt{\dfrac{s_1^2}{n_1} + \dfrac{s_2^2}{n_2}} \rightarrow (61 - 60) \pm 1.796 \sqrt{\dfrac{(3.59)^2}{15} + \dfrac{(2.41)^2}{12}} \rightarrow$

$1 \pm 1.849 \rightarrow -0.849 < \mu_1 - \mu_2 < 2.849$

8.3 TESTING THE DIFFERENCE BETWEEN TWO MEANS (DEPENDENT SAMPLES)

8.3 Try It Yourself Solutions

1a. (1) Independent   (2) Dependent

2.

Before	After	d	d^2
72	73	−1	1
81	80	1	1
76	79	−3	9
74	76	−2	4
75	76	−1	1
80	80	0	0
68	74	−6	36
75	77	−2	4
78	75	3	9
76	74	2	4
74	76	−2	4
77	78	−1	1
		−12	74

a. $H_0: \mu_d \geq 0$ and $H_a: \mu_d < 0$ (claim)

b. $\alpha = 0.05$ and d.f. $= n - 1 = 11$

c. $t_0 \approx -1.796 \rightarrow$ Reject H_0 if $t < -1.796$

d. $\bar{d} = \dfrac{\Sigma d}{n} = \dfrac{-12}{12} = -1$

$s_d = \sqrt{\dfrac{n(\Sigma d^2) - (\Sigma d)^2}{n(n-1)}} = \sqrt{\dfrac{12(74) - (-12)^2}{12(11)}} \approx 2.374$

e. $t = \dfrac{\bar{d} - \mu_d}{\frac{s_d}{\sqrt{n}}} = \dfrac{-1 - 0}{\frac{2.374}{\sqrt{12}}} \approx -1.459$

f. Fail to reject H_0.

g. There is not enough evidence to support the claim.

3.

Before	After	d	d^2
101.8	99.2	2.6	6.76
98.5	98.4	0.1	0.01
98.1	98.2	−0.1	0.01
99.4	99	0.4	0.16
98.9	98.6	0.3	0.09
100.2	99.7	0.5	0.25
97.9	97.8	0.1	0.01
		3.9	7.29

a. $H_0: \mu_d = 0$ and $H_a: \mu_d \neq 0$ (claim)

b. $\alpha = 0.05$ and d.f. $= n - 1 = 6$

c. $t_0 = \pm 2.447 \rightarrow$ Reject H_0 if $t < -2.447$ or $t > 2.447$

d. $\bar{d} = \dfrac{\Sigma d}{n} = \dfrac{3.9}{7} \approx 0.557$

$s_d = \sqrt{\dfrac{n(\Sigma d^2) - (\Sigma d)^2}{n(n-1)}} = \sqrt{\dfrac{7(7.29) - (3.9)^2}{7(6)}} \approx 0.924$

e. $t = \dfrac{\bar{d} - \mu_d}{\frac{s_d}{\sqrt{n}}} = \dfrac{0.557 - 0}{\frac{0.924}{\sqrt{7}}} \approx 1.595$

f. Fail to reject H_0.

g. There is not enough evidence to support the claim.

8.3 EXERCISE SOLUTIONS

1. Two samples are dependent if each member of one sample corresponds to a member of the other sample. Example: The weights of 22 people before starting an exercise program and the weights of the same 22 people six weeks after starting the exercise program. Two samples are independent if the sample selected from one population is not related to the sample selected from the second population. Example: The weights of 25 cats and the weights of 20 dogs.

3. Independent because sample sizes are different and the data represent test scores from different individuals.

5. Dependent because the samples can be paired with respect to each individual.

7. Independent because sample sizes are different and the data represent average speed from different boats.

9. Dependent because the samples can be paired with respect to each car.

11. $H_0: \mu_d \geq 0$ and $H_a: \mu_d < 0$ (claim)

$\alpha = 0.05$ and d.f. $= n - 1 = 9$

$t_0 = -1.833$ (left-tailed)

$$t = \frac{\bar{d} - \mu_d}{\frac{s_d}{\sqrt{n}}} = \frac{10 - 0}{\frac{1.5}{\sqrt{10}}} \approx 21.082$$

Fail to reject H_0. There is not enough evidence to support the claim.

13. $H_0: \mu_d \leq 0$ (claim) and $H_a: \mu_d > 0$

$\alpha = 0.10$ and d.f. $= n - 1 = 15$

$t_0 = 1.341$ (right-tailed)

$$t = \frac{\bar{d} - \mu_d}{\frac{s_d}{\sqrt{n}}} = \frac{6.1 - 0}{\frac{0.36}{\sqrt{16}}} \approx 67.778$$

Reject H_0. There is enough evidence to reject the claim.

15. (a) $H_0: \mu_d \geq 0$ and $H_a: \mu_d < 0$ (claim)

(b) $t_0 = -2.650 \rightarrow$ Reject H_0 if $t < -2.650$.

(c) $\bar{d} \approx -33.714$ and $s_d \approx 42.034$

(d) $t = \dfrac{\bar{d} - \mu_d}{\frac{s_d}{\sqrt{n}}} = \dfrac{-33.714 - 0}{\frac{42.034}{\sqrt{14}}} \approx -3.001$

(e) Reject H_0. There is enough evidence to support the claim that the second SAT scores are improved.

17. (a) $H_0: \mu_d \geq 0$ and $H_a: \mu_d < 0$ (claim)

(b) $t_0 = -1.415 \rightarrow$ Reject H_0 if $t < -1.415$.

(c) $\bar{d} \approx -1.125$ and $s_d \approx 0.871$

(d) $t = \dfrac{\bar{d} - \mu_d}{\frac{s_d}{\sqrt{n}}} = \dfrac{-1.125 - 0}{\frac{0.871}{\sqrt{8}}} \approx -3.653$

(e) Reject H_0. There is enough evidence to support the claim that the fuel additive improved gas mileage.

19. (a) $H_0: \mu_d \leq 0$ and $H_a: \mu_d > 0$ (claim)

(b) $t_0 = 1.796 \rightarrow$ Reject H_0 if $t > 1.796$.

(c) $\bar{d} \approx 16.833$ and $s_d \approx 6.952$

(d) $t = \dfrac{\bar{d} - \mu_d}{\frac{s_d}{\sqrt{n}}} = \dfrac{16.833 - 0}{\frac{6.952}{\sqrt{12}}} \approx 8.388$

(e) Reject H_0. There is enough evidence to support the claim that the new drug reduces systolic blood pressure.

21. (a) H_0: $\mu_d \leq 0$ and H_a: $\mu_d > 0$ (claim)

(b) $t_0 = 2.764 \rightarrow$ Reject H_0 if $t > 2.764$.

(c) $\overline{d} \approx 1.255$ and $s_d \approx 0.441$

(d) $t = \dfrac{\overline{d} - \mu_d}{\frac{s_d}{\sqrt{n}}} = \dfrac{1.255 - 0}{\frac{0.441}{\sqrt{11}}} \approx 9.438$

(e) Reject H_0. There is enough evidence to support the claim that soft tissue therapy and spinal manipulation help reduce the length of time patients suffer from headaches.

23. $\overline{d} \approx -1.467$ and $s_d \approx 0.569$

$$\overline{d} - t_{\alpha/2}\frac{s_d}{\sqrt{n}} < \mu_d < \overline{d} - t_{\alpha/2}\frac{s_d}{\sqrt{n}}$$

$$-1.467 - 1.796\left(\frac{0.569}{\sqrt{12}}\right) < \mu_d < -1.467 + 1.796\left(\frac{0.569}{\sqrt{12}}\right)$$

$$-1.762 < \mu_d < -1.172$$

8.4 TESTING THE DIFFERENCE BETWEEN TWO PROPORTIONS

8.4 Try It Yourself Solutions

1a. H_0: $p_1 = p_2$ and H_a: $p_1 \neq p_2$ (Claim)

b. $\alpha = 0.05$

c. $z_0 = \pm 1.96 \rightarrow$ Reject H_0 if $z < -1.96$ or $z > 1.96$

d. $\overline{p} = \dfrac{x_1 + x_2}{n_1 + n_2} = \dfrac{388 + 236}{7924 + 7364} = 0.041$

$\overline{q} = 0.959$

e. $z = \dfrac{(\hat{p}_1 - \hat{p}_2) - (p_1 - p_2)}{\sqrt{\overline{p}\overline{q}\left(\frac{1}{n_1} + \frac{1}{n_2}\right)}} = \dfrac{(0.049 - 0.032) - (0)}{\sqrt{0.041 \cdot 0.959\left(\frac{1}{7924} + \frac{1}{7364}\right)}} = 5.297$

f. Reject H_0

g. There is enough evidence to support the claim.

2a. H_0: $p_1 \leq p_2$ and H_a: $p_1 > p_2$ (claim)

b. $\alpha = 0.05$

c. $z_0 = 1.645 \rightarrow$ Reject H_0 if $z > 1.645$

d. $\overline{p} = \dfrac{x_1 + x_2}{n_1 + n_2} = \dfrac{1085 + 678}{7924 + 7364} = 0.115$

$\overline{q} = 0.885$

e. $z = \dfrac{(\hat{p}_1 - \hat{p}_2) - (p_1 - p_2)}{\sqrt{\overline{p}\overline{q}\left(\frac{1}{n_1} + \frac{1}{n_2}\right)}} = \dfrac{(0.137 - 0.092) - (0)}{\sqrt{0.115 \cdot 0.885\left(\frac{1}{7924} + \frac{1}{7364}\right)}} = 8.715$

f. Reject H_0

g. There is enough evidence to support the claim.

8.4 EXERCISE SOLUTIONS

1. State the hypotheses and identify the claim. Specify the level of significance. Find the critical value(s) and rejection region(s). Find $\bar{p}$ and $\bar{q}$. Find the standardized test statistic. Make a decision and interpret in the context of the claim.

3. $H_0: p_1 = p_2$ and $H_a: p_1 \neq p_2$ (claim)

$z_0 = \pm 2.575$ (two-tailed test)

$$\bar{p} = \frac{x_1 + x_2}{n_1 + n_2} = \frac{35 + 36}{70 + 60} = 0.546$$

$\bar{q} = 0.454$

$$z = \frac{(\hat{p}_1 - \hat{p}_2) - (p_1 - p_2)}{\sqrt{\bar{p}\bar{q}\left(\frac{1}{n_1} + \frac{1}{n_2}\right)}} = \frac{(0.500 - 0.600) - (0)}{\sqrt{0.546 \cdot 0.454\left(\frac{1}{70} + \frac{1}{60}\right)}} = -1.142$$

Fail to reject H_0. There is not enough evidence to support the claim.

5. $H_0: p_1 \leq p_2$ (claim) and $H_a: p_1 > p_2$

$z_0 = 1.282$ (right-tailed test)

$$\bar{p} = \frac{x_1 + x_2}{n_1 + n_2} = \frac{344 + 304}{860 + 800} = 0.390$$

$\bar{q} = 0.610$

$$z = \frac{(\hat{p}_1 - \hat{p}_2) - (p_1 - p_2)}{\sqrt{\bar{p}\bar{q}\left(\frac{1}{n_1} + \frac{1}{n_2}\right)}} = \frac{(0.400 - 0.380) - (0)}{\sqrt{0.390 \cdot 0.610\left(\frac{1}{860} + \frac{1}{800}\right)}} = 0.835$$

Fail to reject H_0. There is not enough evidence to reject the claim.

7. (a) $H_0: p_1 = p_2$ (claim) and $H_a: p_1 \neq p_2$

(b) $z_0 = \pm 1.96 \rightarrow$ Reject H_0 if $z < -1.96$ or $z > 1.96$

$$\bar{p} = \frac{x_1 + x_2}{n_1 + n_2} = \frac{520 + 865}{1539 + 2055} = 0.385$$

$\bar{q} = 0.615$

(c) $$z = \frac{(\hat{p}_1 - \hat{p}_2) - (p_1 - p_2)}{\sqrt{\bar{p}\bar{q}\left(\frac{1}{n_1} + \frac{1}{n_2}\right)}} = \frac{(0.338 - 0.421) - (0)}{\sqrt{0.385 \cdot 0.615\left(\frac{1}{1539} + \frac{1}{2055}\right)}} = -5.060$$

(d) Reject H_0. There is enough evidence to reject the claim.

9. (a) $H_0: p_1 \geq p_2$ and $H_a: p_1 < p_2$ (claim)

(b) $z_0 = -2.33 \rightarrow$ Reject H_0 if $z < -2.33$

$$\bar{p} = \frac{x_1 + x_2}{n_1 + n_2} = \frac{494 + 574}{2000 + 2000} = 0.267$$

$\bar{q} = 0.733$

(c) $$z = \frac{(\hat{p}_1 - \hat{p}_2) - (p_1 - p_2)}{\sqrt{\bar{p}\bar{q}\left(\frac{1}{n_1} + \frac{1}{n_2}\right)}} = \frac{(0.247 - 0.287) - (0)}{\sqrt{0.267 \cdot 0.733\left(\frac{1}{2000} + \frac{1}{2000}\right)}} = -2.859$$

    (d) Reject H_0. There is enough evidence to reject the claim.

11. (a) $H_0: p_1 = p_2$ (claim) and $H_a: p_1 \neq p_2$

    (b) $z_0 = \pm 1.645 \rightarrow$ Reject H_0 if $z < -1.645$ or $z > 1.645$

$$\bar{p} = \frac{x_1 + x_2}{n_1 + n_2} = \frac{2358 + 2437}{10{,}572 + 8551} = 0.251$$

$$\bar{q} = 0.749$$

    (c) $z = \dfrac{(\hat{p}_1 - \hat{p}_2) - (p_1 - p_2)}{\sqrt{\bar{p}\bar{q}\left(\dfrac{1}{n_1} + \dfrac{1}{n_2}\right)}} = \dfrac{(0.223 - 0.285) - (0)}{\sqrt{0.251 \cdot 0.749\left(\dfrac{1}{10{,}572} + \dfrac{1}{8551}\right)}} = -9.832$

(d) Reject H_0. There is enough evidence to reject the claim.

13. $H_0: p_1 \geq p_2$ and $H_a: p_1 < p_2$ (claim)

$z_0 = -2.33$

$$\bar{p} = \frac{x_1 + x_2}{n_1 + n_2} = \frac{387 + 459}{900 + 1020} = 0.441$$

$$\bar{q} = 0.559$$

$$z = \frac{(\hat{p}_1 - \hat{p}_2) - (p_1 - p_2)}{\sqrt{\bar{p}\bar{q}\left(\dfrac{1}{n_1} + \dfrac{1}{n_2}\right)}} = \frac{(0.43 - 0.45) - (0)}{\sqrt{0.441 \cdot 0.559\left(\dfrac{1}{900} + \dfrac{1}{1020}\right)}} = -0.881$$

Fail to reject H_0. There is not enough evidence to support the claim.

15. $H_0: p_1 = p_2$ (claim) and $H_a: p_1 \neq p_2$

$z_0 = \pm 1.96$

$$\bar{p} = \frac{x_1 + x_2}{n_1 + n_2} = \frac{410 + 363}{1000 + 1100} = 0.368$$

$$\bar{q} = 0.632$$

$$z = \frac{(\hat{p}_1 - \hat{p}_2) - (p_1 - p_2)}{\sqrt{\bar{p}\bar{q}\left(\dfrac{1}{n_1} + \dfrac{1}{n_2}\right)}} = \frac{(0.410 - 0.330) - (0)}{\sqrt{0.368 \cdot 0.632\left(\dfrac{1}{1000} + \dfrac{1}{1100}\right)}} = 3.797$$

Reject H_0. There is enough evidence to reject the claim.

17. $(\hat{p}_1 - \hat{p}_2) \pm z_c \sqrt{\dfrac{\hat{p}_1 \hat{q}_1}{n_1} + \dfrac{\hat{p}_2 \hat{q}_2}{n_2}} \rightarrow (0.117 - 0.088)$

$\pm 1.96 \sqrt{\dfrac{0.117 \cdot 0.883}{977{,}000} + \dfrac{0.088 \cdot 0.912}{1{,}085{,}000}} \rightarrow 0.028 < p_1 - p_2 < 0.030$

CHAPTER 8 REVIEW EXERCISE SOLUTIONS

1. $H_0: \mu_1 \geq \mu_2$ (claim) and $H_1: \mu_1 < \mu_2$

$z_0 = -1.645$

$$z = \frac{(\bar{x}_1 - \bar{x}_2) - (\mu_1 - \mu_2)}{\sqrt{\dfrac{s_1^2}{n_1} + \dfrac{s_2^2}{n_2}}} = \frac{(1.28 - 1.36) - (0)}{\sqrt{\dfrac{(0.28)^2}{76} + \dfrac{(0.23)^2}{65}}} = -.1862$$

Reject H_0. There is enough evidence to reject the claim.

3. $H_0: \mu_1 \geq \mu_2$ and $H_1: \mu_1 < \mu_2$ (claim)

$z_0 = -1.282$

$$z = \frac{(\bar{x}_1 - \bar{x}_2) - (\mu_1 - \mu_2)}{\sqrt{\frac{s_1^2}{n_1} + \frac{s_2^2}{n_2}}} = \frac{(0.28 - 0.33) - (0)}{\sqrt{\frac{(0.11)^2}{41} + \frac{(0.10)^2}{34}}} = -2.060$$

Reject H_0. There is enough evidence to support the claim.

5. $H_0: \mu_1 \geq \mu_2$ and $H_1: \mu_1 < \mu_2$ (claim)

$z_0 = -1.645$

$$z = \frac{(\bar{x}_1 - \bar{x}_2) - (\mu_1 - \mu_2)}{\sqrt{\frac{s_1^2}{n_1} + \frac{s_2^2}{n_2}}} = \frac{(529 - 560) - (0)}{\sqrt{\frac{(43)^2}{36} + \frac{(57)^2}{41}}} \approx -2.713$$

Reject H_0. There is enough evidence to support the claim.

7. $H_0: \mu_1 = \mu_2$ (claim) and $H_a: \mu_1 \neq \mu_2$

d.f. $= n_1 + n_2 - 2 = 31$

$t_0 = \pm 1.96$

$$t = \frac{(\bar{x}_1 - \bar{x}_2) - (\mu_1 - \mu_2)}{\sqrt{\frac{(n_1 - 1)s_1^2 + (n_2 - 1)s_2^2}{n_1 + n_2 - 2}}\sqrt{\frac{1}{n_1} + \frac{1}{n_2}}} = \frac{(250 - 240) - (0)}{\sqrt{\frac{(21 - 1)(26)^2 + (12 - 1)(22)^2}{21 + 12 - 2}}\sqrt{\frac{1}{21} + \frac{1}{12}}} = 1.121$$

Fail to reject H_0. There is not enough evidence to reject the claim.

9. $H_0: \mu_1 \leq \mu_2$ (claim) and $H_a: \mu_1 > \mu_2$

d.f. $= \min\{n_1 - 1, n_2 - 1\} = 24$

$t_0 = 1.711$

$$t = \frac{(\bar{x}_1 - \bar{x}_2) - (\mu_1 - \mu_2)}{\sqrt{\frac{s_1^2}{n_1} + \frac{s_2^2}{n_2}}} = \frac{(183.5 - 184.7) - (0)}{\sqrt{\frac{(1.3)^2}{25} + \frac{(3.9)^2}{25}}} = -1.460$$

Fail to reject H_0. There is not enough evidence to reject the claim.

11. (a) $H_0: \mu_1 \leq \mu_2$ and $H_a: \mu_1 > \mu_2$ (claim)

(b) d.f. $= n_1 + n_2 - 2 = 42$

$t_0 = 1.645$

(c) $\bar{x}_1 = 51.476, s_1 = 11.007, n_1 = 21$

$\bar{x}_2 = 41.522, s_2 = 17.149, n_2 = 23$

$$t = \frac{(\bar{x}_1 - \bar{x}_2) - (\mu_1 - \mu_2)}{\sqrt{\frac{(n_1 - 1)s_1^2 + (n_2 - 1)s_2^2}{n_1 + n_2 - 2}}} = \frac{(51.476 - 41.522) - (0)}{\sqrt{\frac{(21 - 1)(11.007)^2 + (23 - 1)(17.149)^2}{21 + 23 - 2}}\sqrt{\frac{1}{21} + \frac{1}{23}}} = 2.266$$

(d) Reject H_0. There is enough evidence to support the claim.

13. Independent since the two samples of laboratory mice are different groups.

15. $H_0: \mu_d = 0$ (claim) and $H_a: \mu_d \neq 0$

$\alpha = 0.05$ and d.f. $= n - 1 = 99$

$t_0 = \pm 1.96$ (two-tailed test)

$$t = \frac{\bar{d} - \mu_d}{\frac{s_d}{\sqrt{n}}} = \frac{10 - 0}{\frac{12.4}{\sqrt{100}}} = 8.065$$

Reject H_0. There is enough evidence to reject the claim.

17. $H_0: \mu_d \le 6$ (claim) and $H_a: \mu_d > 6$

$\alpha = 0.10$ and d.f. $= n - 1 = 32$

$t_0 = 1.282$ (right-tailed test)

$t = \dfrac{\bar{d} - \mu_d}{\frac{s_d}{\sqrt{n}}} = \dfrac{10.3 - 6}{\frac{1.24}{\sqrt{33}}} = 19.921$

Reject H_0. There is enough evidence to reject the claim.

19. (a) $H_0: \mu_d \le 0$ and $H_a: \mu_d > 0$ (claim)

(b) $t_0 = 1.383 \to$ Reject H_0 if $t > 1.383$

(c) $\bar{d} = 5$ and $s_d \approx 8.743$

(d) $t = \dfrac{\bar{d} - \mu_d}{\frac{s_d}{\sqrt{n}}} = \dfrac{5 - 0}{\frac{8.743}{\sqrt{10}}} \approx 1.808$

(e) Reject H_0.

(f) There is enough evidence to support the claim.

21. $H_0: p_1 = p_2$ and $H_a: p_1 \ne p_2$ (claim)

$z_0 = \pm 1.96$ (two-tailed test)

$\bar{p} = \dfrac{x_1 + x_2}{n_1 + n_2} = \dfrac{375 + 365}{720 + 660} = 0.536$

$\bar{q} = 0.464$

$z = \dfrac{(\hat{p}_1 - \hat{p}_2) - (p_1 - p_2)}{\sqrt{\bar{p}\bar{q}\left(\frac{1}{n_1} + \frac{1}{n_2}\right)}} = \dfrac{(0.521 - 0.553) - (0)}{\sqrt{0.536 \cdot 0.464\left(\frac{1}{720} + \frac{1}{660}\right)}} = -1.198$

Fail to reject H_0. There is not enough evidence to reject the claim.

23. $H_0: p_1 \le p_2$ and $H_a: p_1 > p_2$ (claim)

$z_0 = 1.282$ (right-tailed test)

$\bar{p} = \dfrac{x_1 + x_2}{n_1 + n_2} = \dfrac{227 + 198}{556 + 420} = 0.435$

$\bar{q} = 0.565$

$z = \dfrac{(\hat{p}_1 - \hat{p}_2) - (p_1 - p_2)}{\sqrt{\bar{p}\bar{q}\left(\frac{1}{n_1} + \frac{1}{n_2}\right)}} = \dfrac{(0.408 - 0.471) - (0)}{\sqrt{0.435 \cdot 0.565\left(\frac{1}{556} + \frac{1}{420}\right)}} = -1.971$

Fail to reject H_0. There is not enough evidence to support the claim.

25. (a) $H_0: p_1 = p_2$ (claim) and $H_a: p_1 \ne p_2$

(b) $z_0 = \pm 1.645 \to$ Reject H_0 if $z < -1.645$ or $z > 1.645$

(c) $z = \dfrac{(\hat{p}_1 - \hat{p}_2) - (p_1 - p_2)}{\sqrt{\bar{p}\bar{q}\left(\frac{1}{n_1} + \frac{1}{n_2}\right)}} = \dfrac{(0.110 - 0.133) - (0)}{\sqrt{0.124 \cdot 0.876\left(\frac{1}{200} + \frac{1}{300}\right)}} = -0.776$

(d) Fail to reject H_0.

(e) There is not enough evidence to reject the claim.

CHAPTER 8 QUIZ SOLUTIONS

1. (a) H_0: $\mu_1 \leq \mu_2$ and H_1: $\mu_1 > \mu_2$ (claim)

(b) n_1 and $n_2 > 30 \rightarrow$ Right tailed z-test

(c) $z_0 = 1.645$

(d) $z = \dfrac{(\bar{x}_1 - \bar{x}_2) - (\mu_1 - \mu_2)}{\sqrt{\dfrac{s_1^2}{n_1} + \dfrac{s_2^2}{n_2}}} = \dfrac{(299.5 - 288.9) - (0)}{\sqrt{\dfrac{(2.0)^2}{49} + \dfrac{(1.7)^2}{50}}} = 28.387$

(e) Reject H_0. There is enough evidence to support the claim.

2. (a) H_0: $\mu_1 = \mu_2$ (claim) and H_a: $\mu_1 \neq \mu_2$

(b) $\{n_1$ and $n_2 < 30\}$ and $\{\sigma^1$ and σ^2 are unknown$\} \rightarrow$ Two tailed t-test (assume vars are equal)

(c) d.f. $= n_1 + n_2 - 2 = 26$

$t_0 = \pm 2.779$

(d) $t = \dfrac{(\bar{x}_1 - \bar{x}_2) - (\mu_1 - \mu_2)}{\sqrt{\dfrac{(n_1 - 1)s_1^2 + (n_2 - 1)s_2^2}{n_1 + n_2 - 2}}} = \dfrac{(232.2 - 230) - (0)}{\sqrt{\dfrac{(13 - 1)(1.3)^2 + (15 - 1)(1.4)^2}{13 + 15 - 2}}\sqrt{\dfrac{1}{13} + \dfrac{1}{15}}} = 4.285$

(e) Reject H_0. There is enough evidence to reject the claim.

3. (a) H_0: $p_1 \leq p_2$ and H_a: $p_1 > p_2$ (claim)

(b) Testing 2 proportions $\rightarrow$ Right tailed z-test

(c) $z_0 = 1.282$

(d) $\bar{p} = \dfrac{x_1 + x_2}{n_1 + n_2} = \dfrac{64,800 + 8560}{1,296,000 + 856,000} = 0.034$

$\bar{q} = 0.966$

$z = \dfrac{(\hat{p}_1 - \hat{p}_2) - (p_1 - p_2)}{\sqrt{\bar{p}\bar{q}\left(\dfrac{1}{n_1} + \dfrac{1}{n_2}\right)}} = \dfrac{(0.05 - 0.01) - (0)}{\sqrt{0.034 \cdot 0.966\left(\dfrac{1}{1,296,000} + \dfrac{1}{856,000}\right)}} \approx 158.471$

(e) Reject H_0. There is enough evidence to support the claim.

4. (a) H_0: $\mu_d \leq 0$ and H_a: $\mu_d > 0$ (claim)

(b) Dependent samples $\rightarrow$ Dependent t-test

(c) $t_0 = 1.796$

(d) $t = \dfrac{\bar{d} - \mu_d}{\dfrac{s_d}{\sqrt{n}}} = \dfrac{68.5 - 0}{\dfrac{26.318}{\sqrt{12}}} = 9.016$

(e) Reject H_0. There is enough evidence to support the claim.

Correlation and Regression

9.1 CORRELATION

9.1 Try It Yourself Solutions

1ab.

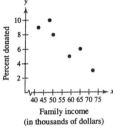

Family income
(in thousands of dollars)

   c. Yes, it appears that there is a negative linear correlation. As family income increases, the percent of income donated to charity decreases.

2ab.

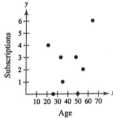

Age

   c. No, it appears that there is no correlation between age and subscriptions.

3ab.

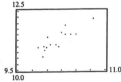

   c. Yes, there appears to be a positive linear relationship between men's winning time and women's winning time.

4a. $n = 6$

b.

x	y	xy	x^2	y^2
42	9	378	1764	81
48	10	480	2304	100
50	8	400	2500	64
59	5	295	3481	25
65	6	390	4225	36
72	3	216	5184	9
$\Sigma x = 336$	$\Sigma y = 41$	$\Sigma xy = 2159$	$\Sigma x^2 = 19458$	$\Sigma y^2 = 315$

   c. $r = \dfrac{n\Sigma xy - (\Sigma x)(\Sigma y)}{\sqrt{n\Sigma x^2 - (\Sigma x)^2}\sqrt{n\Sigma y^2 - (\Sigma y)^2}} = \dfrac{6(2159) - (336)(41)}{\sqrt{6(19{,}458) - (336)^2}\sqrt{6(315) - (41)^2}} \approx -0.916$

   d. Since r is close to -1, there appears to be a strong negative linear correlation between income level and donating percent.

5a. Enter data

b. $r \approx 0.832$

c. Since r is close to 1, there appears to be a strong positive linear correlation between men's winning time and women's winning time.

6a. $H_0: \rho = 0$ and $H_a: \rho \neq 0$

b. $\alpha = 0.01$

c. d.f. $= n - 2 = 33$

d. $\pm 2.576 \rightarrow$ Reject H_0 if $t < -2.576$ or $t > 2.576$

e. $t = \dfrac{r}{\sqrt{\dfrac{1 - r^2}{n - 2}}} = \dfrac{0.970}{\sqrt{\dfrac{1 - (0.970)^2}{35 - 2}}} \approx 22.921$

f. Reject H_0.

g. There is enough evidence in the sample to conclude that a significant correlation exists.

9.1 EXERCISE SOLUTIONS

1. Positive linear correlation

3. No linear correlation (but there is a nonlinear correlation between the variables)

5. (c) As age increases, income also increases.

7. (b) As age increases, the balance on student loans decreases.

9. Explanatory variable: Amount of water consumed.
Response variable: Weight loss.

11. (a)

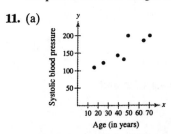

(b)

x	y	xy	x^2	y^2
16	109	1744	256	11,881
25	122	3050	625	14,884
39	143	5577	1521	20,449
45	132	5940	2025	17,424
49	199	9751	2401	39,601
64	185	11,840	4096	34,225
70	199	13,930	4900	39,601
$\Sigma x = 308$	$\Sigma y = 1089$	$\Sigma xy = 51,832$	$\Sigma x^2 = 15,824$	$\Sigma y^2 = 178,065$

$$r = \frac{n\Sigma xy - (\Sigma x)(\Sigma y)}{\sqrt{n\Sigma x^2 - (\Sigma x)^2}\sqrt{n\Sigma y^2 - (\Sigma y)^2}} = \frac{7(51,832) - (308)(1089)}{\sqrt{7(15,824) - (308)^2}\sqrt{7(178,065) - (1089)^2}} \approx 0.883$$

(c) Strong positive linear correlation

13. (a)

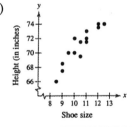

Hours studying

(b)

x	y	xy	x²	y²
0	40	0	0	1600
1	41	41	1	1681
2	51	102	4	2601
4	48	192	16	2304
4	64	256	16	4096
5	69	345	25	4761
5	73	365	25	5329
5	75	375	25	5625
6	68	408	36	4624
6	93	558	36	8649
7	84	588	49	7056
7	90	630	49	8100
8	95	760	64	9025
$\Sigma x = 60$	$\Sigma y = 891$	$\Sigma xy = 4620$	$\Sigma x^2 = 346$	$\Sigma y^2 = 65{,}451$

$$r = \frac{n\Sigma xy - (\Sigma x)(\Sigma y)}{\sqrt{n\Sigma x^2 - (\Sigma x)^2}\sqrt{n\Sigma y^2 - (\Sigma y)^2}} = \frac{13(4620) - (60)(891)}{\sqrt{13(346) - (60)^2}\sqrt{13(65{,}451) - (891)^2}} \approx 0.923$$

(c) Strong positive linear correlation

15. (a)

Height (in inches)

Shoe size

(b)

x	y	xy	x²	y²
8.5	66.0	561.00	72.25	4356.00
9.0	68.5	616.50	81.00	4692.25
9.0	67.5	607.50	81.00	4556.25
9.5	70.0	665.00	90.25	4900.00
10.0	70.0	700.00	100.00	4900.00
10.0	72.0	720.00	100.00	5184.00
10.5	71.5	750.75	110.25	5112.25
10.5	69.5	729.75	110.25	4830.25
11.0	71.5	786.50	121.00	5112.25
11.0	72.0	792.00	121.00	5184.00
11.0	73.0	803.00	121.00	5329.00
12.0	73.5	882.00	144.00	5402.25
12.0	74.0	888.00	144.00	5476.00
12.5	74.0	925.00	156.25	5476.00
$\Sigma x = 146.5$	$\Sigma y = 993.0$	$\Sigma xy = 10{,}427.0$	$\Sigma x^2 = 1552.3$	$\Sigma y^2 = 70{,}510.5$

$$r = \frac{n\Sigma xy - (\Sigma x)(\Sigma y)}{\sqrt{n\Sigma x^2 - (\Sigma x)^2}\sqrt{n\Sigma y^2 - (\Sigma y)^2}} = \frac{14(10{,}427.0) - (146.5)(993.0)}{\sqrt{14(1552.3) - (146.5)^2}\sqrt{14(70{,}510.5) - (993.0)^2}} \approx 0.926$$

(c) Strong positive linear correlation

17. (a)

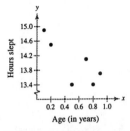

Age (in years)

(b)

x	y	xy	x^2	y^2
0.1	14.9	1.49	0.01	222.01
0.2	14.5	2.90	0.04	210.25
0.5	13.4	6.70	0.25	179.56
0.7	14.1	9.87	0.49	198.81
0.8	13.4	10.72	0.64	179.56
0.9	13.7	12.33	0.81	187.69
$\Sigma x = 3.2$	$\Sigma y = 84$	$\Sigma xy = 44.01$	$\Sigma x^2 = 2.24$	$\Sigma y^2 = 1177.88$

$$r = \frac{n\Sigma xy - (\Sigma x)(\Sigma y)}{\sqrt{n\Sigma x^2 - (\Sigma x)^2}\sqrt{n\Sigma y^2 - (\Sigma y)^2}} = \frac{6(44.01) - (3.2)(84)}{\sqrt{6(2.24) - (3.2)^2}\sqrt{6(1177.88) - (84)^2}} \approx -0.789$$

(c) Negative linear correlation

19. $r = -0.84$ represents a stronger correlation since it is closer to -1 than $r = 0.73$ is to $+1$.

21. State the null and alternative hypotheses. Specify the level of significance and determine the degrees of freedom. Identify the rejection regions and calculate the standardized test statistic. Make a decision and interpret in the context of the original claim.

23. H_0: $\rho = 0$ and H_a: $\rho \neq 0$

$\alpha = 0.05$

d.f. $= n - 2 = 5$

$\pm 2.571 \rightarrow$ Reject H_0 if $t < -2.571$ or $t > 2.571$

$$t = \frac{r}{\sqrt{\dfrac{1 - r^2}{n - 2}}} = \frac{0.50}{\sqrt{\dfrac{1 - (0.50)^2}{7 - 2}}} \approx 1.291$$

Fail to reject H_0. There is not enough evidence to support the claim a significant linear correlation exists.

25. H_0: $\rho = 0$ and H_a: $\rho \neq 0$

$\alpha = 0.01$

d.f. $= n - 2 = 23$

$\pm 2.807 \rightarrow$ Reject H_0 if $t < -2.807$ or $t > 2.807$

$$t = \frac{r}{\sqrt{\dfrac{1 - r^2}{n - 2}}} = \frac{-0.83}{\sqrt{\dfrac{1 - (-0.83)^2}{25 - 2}}} \approx -7.137$$

Reject H_0. There is enough evidence to conclude that a significant linear correlation exists.

27. (a) H_0: $\rho = 0$ and H_a: $\rho \neq 0$

(b) $\alpha = 0.10$

d.f. $= n - 2 = 11$

$\pm 1.796 \rightarrow$ Reject H_0 if $t < -1.796$ or $t > 1.796$

(c) $r \approx 0.923$

$$t = \frac{r}{\sqrt{\dfrac{1-r^2}{n-2}}} = \frac{0.923}{\sqrt{\dfrac{1-(0.923)^2}{13-2}}} \approx 7.955$$

(d) Reject H_0. There is enough evidence to conclude that a significant linear correlation exists.

29. (a) $H_0: \rho = 0$ and $H_a: \rho \neq 0$

(b) $\alpha = 0.05$ and d.f. $= n - 2 = 12$

$\pm 2.179 \rightarrow$ Reject H_0 if $t < -2.179$ or $t > 2.179$

(c) $r \approx 0.926$

$$t = \frac{r}{\sqrt{\dfrac{1-r^2}{n-2}}} = \frac{0.926}{\sqrt{\dfrac{1-(0.926)^2}{14-2}}} \approx 8.497$$

(d) Reject H_0. There is enough evidence to conclude that a significant linear correlation exists.

31. The correlation coefficient remains unchanged when the *x*-values and *y*-values are switched.

33. Answers will vary.

9.2 LINEAR REGRESSION

9.2 Try It Yourself Solutions

1a. $n = 6$

x	y	xy	x^2
42	9	378	1764
48	10	480	2304
50	8	400	2500
59	5	295	3481
65	6	390	4225
72	3	216	5184
$\Sigma x = 336$	$\Sigma y = 41$	$\Sigma xy = 2159$	$\Sigma x^2 = 19458$

b. $m = \dfrac{n\Sigma xy - (\Sigma x)(\Sigma y)}{n\Sigma x^2 - (\Sigma x)^2} = \dfrac{6(2159) - (336)(41)}{6(19,458) - (336)^2} \approx -0.2134$

c. $b = \bar{y} - m\bar{x} = \left(\dfrac{41}{6}\right) - (-0.2134)\left(\dfrac{336}{6}\right) \approx 18.7837$

d. $\hat{y} = -0.213x + 18.784$

2a. Enter the data.

b. $m \approx 1.494; \ b \approx -3.909$

c. $\hat{y} = 1.494x - 3.909$

3ab. (1) $\hat{y} = 11.824(2) + 35.301 \approx 58.949$

(2) $\hat{y} = 11.824(3.32) + 35.301 \approx 74.557$

c. (1) 58.949 minutes

  (2) 74.557 minutes

9.2 EXERCISE SOLUTIONS

1. c

3. d

5.

x	y	xy	x^2
70	8.3	581.0	4900
72	10.5	756.0	5184
75	11.0	825.0	5625
76	11.4	866.4	5776
85	12.9	1096.5	7225
78	14.0	1092.0	6084
77	16.3	1255.1	5929
80	18.0	1440.0	6400
$\Sigma x = 613$	$\Sigma y = 102.4$	$\Sigma xy = 7912$	$\Sigma x^2 = 47{,}123$

$$m = \frac{n\Sigma xy - (\Sigma x)(\Sigma y)}{n\Sigma x^2 - (\Sigma x)^2} = \frac{8(7912) - (613)(102.4)}{8(47{,}123) - (613)^2} \approx 0.43193$$

$$b = \bar{y} - m\bar{x} = \left(\frac{102.4}{8}\right) - (0.43193)\left(\frac{613}{8}\right) \approx 20.2966$$

$\hat{y} = 0.4319x - 20.297$ best fits the data.

7. c

9. a

11.

x	y	xy	x^2
16	109	1744	256
25	122	3050	625
39	143	5577	1521
45	132	5940	2025
49	199	9751	2401
64	185	11,840	4096
70	199	13,930	4900
$\Sigma x = 308$	$\Sigma y = 1089$	$\Sigma xy = 51{,}832$	$\Sigma x^2 = 15{,}824$

$$m = \frac{n\Sigma xy - (\Sigma x)(\Sigma y)}{n\Sigma x^2 - (\Sigma x)^2} = \frac{7(51{,}832) - (308)(1089)}{7(15{,}824) - (308)^2} \approx 1.724$$

$$b = \bar{y} - m\bar{x} = \left(\frac{1089}{7}\right) - (1.724)\left(\frac{308}{7}\right) \approx 79.733$$

$\hat{y} = 1.724x + 79.733$

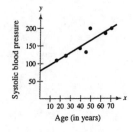

(a) $\hat{y} = 1.724(18) + 79.733 \approx 110$

(b) $\hat{y} = 1.724(71) + 79.733 \approx 202$

(c) $\hat{y} = 1.724(29) + 79.733 \approx 129$

(d) $\hat{y} = 1.724(55) + 79.733 \approx 175$

13.

x	y	xy	x^2
0	40	0	0
1	41	41	1
2	51	102	4
4	48	192	16
4	64	256	16
5	69	345	25
5	73	365	25
5	75	375	25
6	68	408	36
6	93	558	36
7	84	588	49
7	90	630	49
8	95	760	64
60	891	4620	346

$$m = \frac{n\Sigma xy - (\Sigma x)(\Sigma y)}{n\Sigma x^2 - (\Sigma x)^2} = \frac{13(4620) - (60)(891)}{13(346) - (60)^2} \approx 7.350$$

$$b = \bar{y} - m\bar{x} = \left(\frac{60}{13}\right) - (7.350)\left(\frac{891}{13}\right) \approx 34.617$$

$$\hat{y} = 7.350x + 34.617$$

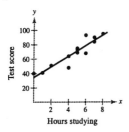

(a) $\hat{y} = 7.350(3) + 34.617 \approx 56.7$

(b) $\hat{y} = 7.350(6.5) + 34.617 \approx 82.4$

(c) It is not meaningful to predict the value of y for $x = 13$ because $x = 13$ is outside the range of the original data.

(d) $\hat{y} = 7.350(4.5) + 34.617 \approx 67.7$

15.

x	y	xy	x^2
8.5	66.0	561.00	72.25
9.0	68.5	616.50	81.00
'9.0	67.5	607.50	81.00
9.5	70.0	665.00	90.25
10.0	70.0	700.00	100.00
10.0	72.0	720.00	100.00
10.5	71.5	750.75	110.25
10.5	69.5	729.75	110.25
11.0	71.5	786.50	121.00
11.0	72.0	792.00	121.00
11.0	73.0	803.00	121.00
12.0	73.5	882.00	144.00
12.0	74.0	888.00	144.00
12.5	74.0	925.00	156.25
146.5	993.0	10,427.0	1552.3

$$m = \frac{n\Sigma xy - (\Sigma x)(\Sigma y)}{n\Sigma x^2 - (\Sigma x)^2} = \frac{14(10,427.0) - (146.5)(993.0)}{14(1552.3) - (146.5)^2} \approx 1.870$$

$$b = \bar{y} - m\bar{x} = \left(\frac{993.0}{14}\right) - (1.870)\left(\frac{146.5}{14}\right) \approx 51.360$$

$$\hat{y} = 1.870x + 51.360$$

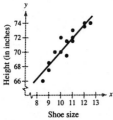

Shoe size

(a) $\hat{y} = 1.870(11.5) + 51.360 \approx 72.865$

(b) $\hat{y} = 1.870(11.5) + 51.360 \approx 66.320$

(c) It is not meaningful to predict the value of y for $x = 15.5$ because $x = 15.5$ is outside the range of the original data.

(d) $\hat{y} = 1.870(11.5) + 51.360 \approx 70.060$

17. Substitute a value x into the equation of a regression line and solve for y.

19. (a) $\hat{y} = 1.724x + 79.733$

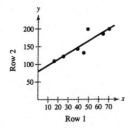

Row 1

(b) $\hat{y} = 0.453x - 26.448$

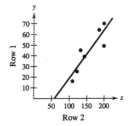

(c) The slope of the line keeps the same sign, but the values of m and b change.

21. $\hat{y} = -1.143x + 89.256$

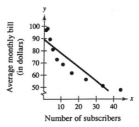

23. $r \approx -0.906$

$H_0: \rho = 0$ and $H_a: \rho \neq 0$

Critical values: $t_0 = \pm 3.355$

$$t = \frac{r}{\sqrt{\dfrac{1-r^2}{n-2}}} = \frac{-0.906}{\sqrt{\dfrac{1-(-0.906)^2}{10-2}}} \approx -6.054$$

Reject H_0. There is enough evidence to conclude that a significant linear correlation exists.

25. Answers will vary.

9.3 MEASURES OF REGRESSION AND PREDICTION INTERVALS

9.3 Try It Yourself Solutions

1a. $r = 0.970$

b. $r^2 = (0.970)^2 = 0.941$

c. 94.1% of the variation in the times is explained.

5.9% of the variation is unexplained.

2a.

x_i	y_i	$\hat{y}_i$	$(y_i - \hat{y}_i)^2$
15	26	28.392	5.721
20	32	35.419	11.689
20	38	35.419	6.662
30	56	49.473	42.602
40	54	63.527	90.764
45	78	70.554	55.442
50	80	77.581	5.851
60	88	91.635	13.214
			$\Sigma = 231.946$

b. $n = 8$

c. $s_e = \sqrt{\dfrac{\Sigma(y_i - \hat{y}_i)^2}{n-2}} = \sqrt{\dfrac{231.946}{6}} \approx 6.218$

d. The standard deviation of the weekly sales for a specific radio ad time is about $621.80.

3a. $n = 8$, d.f. = 6, $t_c = 2.447$, $s_e \approx 10.290$

b. $\hat{y} = 50.729x + 104.061 = 50.729(2.5) + 104.061 \approx 230.884$

c. $E = t_c s_e \sqrt{1 + \dfrac{1}{n} + \dfrac{n(x - \bar{x})^2}{n(\Sigma x^2) - (\Sigma x)^2}} = (2.447)(10.290)\sqrt{1 + \dfrac{1}{8} + \dfrac{8(2.5 - 1.975)^2}{8(32.44) - (15.8)^2}} \approx 29.236$

d. $\hat{y} \pm E \rightarrow (201.648, 260.120)$

e. You can be 95% confident that the company sales will be between $201,648 and $260,120 when advertising expenditures are $2500.

9.3 EXERCISE SOLUTIONS

1. Total variation = $\Sigma(y_i - \bar{y})^2$; the sum of the squares of the differences between the y-values of each ordered pair and the mean of the y-values of the ordered pairs.

3. Unexplained variation = $\Sigma(y_i - \hat{y}_i)^2$; the sum of the squares of the differences between the observed y-values and the predicted y-values.

5. $r^2 = (0.250)^2 \approx 0.063$

6.3% of the variation is explained. 93.7% of the variation is unexplained.

7. $r = (-0.891)^2 \approx 0.794$

79.4% of the variation is explained. 20.6% of the variation is unexplained.

9. (a) $r^2 = \dfrac{\Sigma(\hat{y}_i - \bar{y})^2}{\Sigma(y_i - \bar{y})^2} \approx 0.817$

81.7% of the variation in proceeds can be explained by the variation in the number of issues and 18.3% of the variation is unexplained.

(b) $s_e = \sqrt{\dfrac{\Sigma(y_i - \hat{y}_i)^2}{n-2}} = \sqrt{\dfrac{363,597,812.496}{10}} \approx 6029.907$

The standard deviation of the proceeds for a specific number of issues is about $6,029,907,000.

11. (a) $r^2 = \dfrac{\Sigma(\hat{y}_i - \bar{y})^2}{\Sigma(y_i - \bar{y})^2} \approx 0.985$

98.5% of the variation in sales can be explained by the variation in the total square footage and 1.5% of the variation is unexplained.

(b) $s_e = \sqrt{\dfrac{\Sigma(y_i - \hat{y}_i)^2}{n-2}} = \sqrt{\dfrac{11,440.051}{9}} \approx 35.652$

The standard deviation of the sales for a specific total square footage is about 35,652,000,000.

13. (a) $r^2 = \dfrac{\Sigma(\hat{y}_i - \bar{y})^2}{\Sigma(y_i - \bar{y})^2} \approx 0.998$

99.8% of the variation in the median weekly earnings of female workers can be explained by the variation in the median weekly earnings of male workers and 0.2% of the variation is unexplained.

(b) $s_e = \sqrt{\dfrac{\Sigma(y_i - \hat{y}_i)^2}{n - 2}} = \sqrt{\dfrac{79.628}{3}} \approx 5.147$

The standard deviation of the median weekly earnings of female workers for a specific median weekly earnings of male workers is about $5.147.

15. (a) $r^2 = \dfrac{\Sigma(\hat{y}_i - \bar{y})^2}{\Sigma(y_i - \bar{y})^2} \approx 0.992$

99.2% of the variation in the money spent can be explained by the variation in the money raised and 0.8% of the variation is unexplained.

(b) $s_e = \sqrt{\dfrac{\Sigma(y_i - \hat{y}_i)^2}{n - 2}} = \sqrt{\dfrac{1{,}551.176}{6}} \approx 16.079$

The standard deviation of the money spent for a specified amount of money raised is about $16,079,000.

17. $n = 12$, d.f. $= 10$, $t_c = 2.228$, $s_e = 6029.907$

$\hat{y} = 55.88x - 7189.033 = 55.884(712) - 7189.033 \approx 32{,}600.375$

$E = t_c s_e \sqrt{1 + \dfrac{1}{n} + \dfrac{n(x - \bar{x})^2}{n(\Sigma x^2) - (\Sigma x)^2}} = (2.228)(6029.907)\sqrt{1 + \dfrac{1}{12} + \dfrac{12(712 - 5703/12)^2}{12(3{,}228{,}613) - (5703)^2}} \approx 14{,}664.591$

$\hat{y} \pm E \rightarrow (17{,}935.784, 47{,}264.966)$

You can be 95% confident that the proceeds will be between 17,935,784,000 and $47,264,966,000 when the number of initial offerings is 712.

19. $n = 11$, d.f. $= 9$, $t_c = 1.833$, $s_e \approx 35.652$

$\hat{y} = 230.8x - 289.8 = 230.8(4.5) - 289.8 \approx 748.800$

$E = t_c s_e \sqrt{1 + \dfrac{1}{n} + \dfrac{n(x - \bar{x})^2}{n(\Sigma x^2) - (\Sigma x)^2}} = (1.833)(35.652)\sqrt{1 + \dfrac{1}{11} + \dfrac{11(4.5 - 43.3/11)^2}{11(184.93) - (43.3)^2}} \approx 68.939$

$\hat{y} \pm E \rightarrow (679.861, 817.739)$

You can be 90% confident that the sales will be between $679,861,000,000 and $817,739,000,000 when the total square footage is 4.5 billion.

21. $n = 5$, d.f. $= 3$, $t_c = 5.841$, $s_e \approx 5.147$

$\hat{y} = 0.898x - 82.291 = 0.898(500) - 82.291 \approx 366.709$

$E = t_c s_e \sqrt{1 + \dfrac{1}{n} + \dfrac{n(x - \bar{x})^2}{n(\Sigma x^2) - (\Sigma x)^2}} = (5.841)(5.147)\sqrt{1 + \dfrac{1}{5} + \dfrac{5(500 - 2311/5)^2}{5(1{,}107{,}823) - (2311)^2}} \approx 33.424$

$\hat{y} \pm E \rightarrow (333.285, 400.133)$

You can be 99% confident that the median earnings of female workers will be between $333.285 and $400.133 when the median weekly earnings of male workers is $500.

23. $n = 8$, d.f. $= 6$, $t_c = 2.306$, $s_e \approx 16.079$

$\hat{y} = 1.020x - 25.854 = 1.020(775.8) - 25.854 \approx 765.462$

$E = t_c s_e \sqrt{1 + \dfrac{1}{n} + \dfrac{n(x - \bar{x})^2}{n(\Sigma x^2) - (\Sigma x)^2}} = (2.306)(16.079)\sqrt{1 + \dfrac{1}{8} + \dfrac{8(775.8 - 4363.5/8)^2}{8(2{,}564{,}874) - (4363.5)^2}} \approx 44.060$

$\hat{y} \pm E \rightarrow (721.402, 809.522)$

You can be 95% confident that the money spent in congressional campaigns will be between $721.402 million and $809.522 million when the money raised is $775.8 million.

25. $\bar{y} = 6.786$

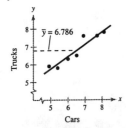

27.

x_i	y_i	$\hat{y}_i$	$\hat{y}_i - \hat{y}$	$y_i - \hat{y}_i$	$y_i - \hat{y}$
8.1	7.8	7.893	1.107	-0.093	1.014
7.7	7.6	7.616	0.830	-0.016	0.814
6.5	6.5	6.785	-0.002	-0.285	-0.286
6.9	7.6	7.062	0.276	0.538	0.814
6.0	6.3	6.438	-0.348	-0.138	-0.486
5.4	5.8	6.022	-0.764	-0.222	-0.986
4.9	5.9	5.676	-1.11	0.224	-0.886

29. $r^2 \approx 0.887$

31. $\hat{y} = 0.693x + 2.280 = 0.693(7.3) + 2.280 \approx 7.339$

$$E = t_c s_e \sqrt{1 + \frac{1}{n} + \frac{n(x - \bar{x})^2}{n(\Sigma x^2) - (\Sigma x)^2}} = (2.571)(0.316)\sqrt{1 + \frac{1}{7} + \frac{7(7.3 - 45.5/7)^2}{7(303.930) - (45.5)^2}} \approx 0.898$$

$\hat{y} \pm E \rightarrow (6.441, 8.237)$

9.4 MULTIPLE REGRESSION

9.4 Try It Yourself Solutions

1a. Enter data.

  b. $\hat{y} = 46.385 + 0.540x_1 - 4.897x_2$

  2. (1) $\hat{y} = 46.385 + 0.540(89) - 4.897(1) = 89.548$

     (2) $\hat{y} = 46.385 + 0.540(78) - 4.897(3) = 73.814$

9.4 EXERCISE SOLUTIONS

  1. $\hat{y} = 6503 - 14.8x_1 + 12.2x_2$

    (a) $\hat{y} = 6503 - 14.8(1458) + 12.2(1450) = 2614.6$

    (b) $\hat{y} = 6503 - 14.8(1500) + 12.2(1475) = 2298$

    (c) $\hat{y} = 6503 - 14.8(1400) + 12.2(1385) = 2680$

    (d) $\hat{y} = 6503 - 14.8(1525) + 12.2(1500) = 2233$

  3. $\hat{y} = -52.2 + 0.3x_1 + 4.5x_2$

    (a) $\hat{y} = -52.2 + 0.3(70) + 4.5(8.6) = 7.5$

    (b) $\hat{y} = -52.2 + 0.3(65) + 4.5(11.0) = 16.8$

(c) $\hat{y} = -52.2 + 0.3(83) + 4.5(17.6) = 51.9$

(d) $\hat{y} = -52.2 + 0.3(87) + 4.5(19.6) = 62.1$

5. $\hat{y} = -256.293 + 103.502x_1 + 14.649x_2$

(a) $s = 34.16$

(b) $r^2 = 0.988$

(c) The standard deviation of the predicted sales given a specific total square footage and number of shopping centers is \$34.16 billion. The multiple regression model explains 98.8% of the variation in y.

7. $n = 11, k = 2, r = 0.988$

$$r_{adj}^2 = 1 - \left[\frac{(1 - r^2)(n - 1)}{n - k - 1} \right] = 0.985$$

CHAPTER 9 REVIEW EXERCISE SOLUTIONS

1.

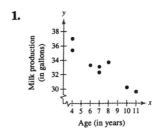

Age (in years)

$r \approx -0.939$; negative linear correlation; milk production decreases with age.

3. $H_0: \rho = 0$ and $H_a: \rho \neq 0$

$\alpha = 0.10$, d.f. $= n - 2 = 24$

Critical value is ± 1.711

$$t = \frac{r}{\sqrt{\dfrac{1 - r^2}{n - 2}}} = \frac{0.24}{\sqrt{\dfrac{1 - (0.24)^2}{26 - 2}}} = 1.211$$

Fail to reject H_0. There is not enough evidence to conclude that a significant linear correlation exists.

5. $H_0: \rho = 0$ and $H_a: \rho \neq 0$

$\alpha = 0.05$, d.f. $= n - 2 = 6$

Critical value is ± 2.447

$$t = \frac{r}{\sqrt{\dfrac{1 - r^2}{n - 2}}} = \frac{-0.939}{\sqrt{\dfrac{1 - (-0.939)^2}{8 - 2}}} = -6.688$$

Reject H_0. There is enough evidence to conclude that a significant linear correlation exists.

7. $\hat{y} = 0.679x + 26.345$

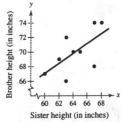

$r \approx 0.625$ (Moderate positive linear correlation)

9. (a) $\hat{y} = 0.679(61) + 26.345 = 67.764$

  (b) $\hat{y} = 0.679(66) + 26.345 = 71.159$

  (c) (not meaningful since $x = 71$ in. is outside range of data)

11. $r^2 = (-0.553)^2 = 0.306$

13. $r^2 = (0.181)^2 = 0.033$

15. (a) $r^2 = 0.897$

  89.7% of the variation in y is explained by the model.

  (b) $s_e = 568.0$

  The standard error of the cooling capacity for a specific living area is 568.0 BTU/hr.

17. $\hat{y} = 0.679(64) + 26.345$

  $= 69.801$

$$E = t_c s_e \sqrt{1 + \frac{1}{n} + \frac{n(x - \bar{x})^2}{n\Sigma x^2 - (\Sigma x)^2}}$$

$$\approx (1.895)(2.397)\sqrt{1 + \frac{1}{9} + \frac{9(64 - 643)^2}{9(37,305) - 579^2}}$$

  ≈ 4.791

$\hat{y} - E < y < \hat{y} + E$

$69.801 - 4.791 < y < 69.801 + 4.791$

$65.01 < y < 74.592$

19. $\hat{y} = 0.3003.0 + 9.468(720)$

  $= 9819.96$

$$E = t_c s_e \sqrt{1 + \frac{1}{n} + \frac{n(x - \bar{x})^2}{n\Sigma x^2 - (\Sigma x)^2}}$$

$$= (2.447)(568)\sqrt{1 + \frac{1}{8} + \frac{8(720 - 499.4)^2}{8(2,182,275) - 3995^2}}$$

  ≈ 1635.63

$\hat{y} - E < y < \hat{y} + E$

$9819.96 - 1635.63 < y < 9819.96 + 1635.63$

$8184.33 < y < 11,455.59$

21. $\hat{y} = 6.317 + 0.8217x_1 + 0.031x_2 - 0.004x_3$

23. (a) 21.705   (b) 25.210   (c) 30.100   (d) 25.860

CHAPTER 9 QUIZ SOLUTIONS

1.

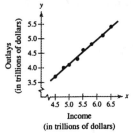

The data appear to have a positive correlation. The outlays increase as the incomes increase.

2. $r \approx 0.997 \rightarrow$ Strong positive linear correlation

3. $H_0: \rho = 0$ and $H_a: \rho \neq 0$

$\alpha = 0.05$, d.f. $= n - 2 = 6$

Critical value is ± 2.447

$$t = \frac{r}{\sqrt{\dfrac{1 - r^2}{n - 2}}} = \frac{0.997}{\sqrt{\dfrac{1 - (0.997)^2}{8 - 2}}} = 31.552$$

Reject H_0. There is enough evidence to conclude that a significant correlation exists.

4. $\hat{y} = 0.838x - 0.069$

5. $\hat{y} = 0.838(5.3) - 0.069 = 4.372$

6. $r^2 = 0.995 \rightarrow 99.5\%$ of the variation in y is explained by the regression model.

7. $s_e = 0.046$

The standard deviation of personal outlays for a specified personal income is $0.046 trillion.

8. $\hat{y} = 0.838(6.4) - 0.069$

$\quad = 5.294$

$$E = t_c s_e \sqrt{1 + \frac{1}{n} + \frac{n(x - \bar{x})^2}{n\Sigma x^2 - (\Sigma x)^2}}$$

$$\quad = 2.447(0.046)\sqrt{1 + \frac{1}{8} + \frac{8(6.4 - 5.45)^2}{8(240.96) - 43.6^2}}$$

$\quad \approx 0.133$

$\hat{y} - E < y < \hat{y} + E$

$5.294 - 0.133 < y < 5.294 + 0.133$

$5.16 < y < 5.43$

You can be 95% confident that the personal outlays will be between $5.16 trillion and $5.43 trillion when personal income is $6.4 trillion.

9. (a) 1311.150   (b) 961.110   (c) 1120.900

(d) 1386.740; x_2 has the greatest influence on y.

CUMULATIVE TEST SOLUTIONS FOR CHAPTERS 7–9

1. $H_0: \mu \geq 20.1$ (claim) and $H_a: \mu < 20.1$

$z_0 = -2.33$

$$z = \frac{\bar{x} - \mu}{\dfrac{s}{\sqrt{n}}} = \frac{18.9 - 20.1}{\dfrac{6.7}{\sqrt{73}}} = -1.530$$

Fail to reject H_0. There is not enough evidence to reject the claim.

2. Type I error will occur if H_0 is rejected when $\mu \geq 20.1$. Type II error will occur if H_0 is not rejected when $\mu < 20.1$.

3. $H_0: \mu_1 - \mu_2 = 0$ (claim) and $H_a: \mu_1 - \mu_2 \neq 0$

$z_0 = \pm 1.96$

$$z = \frac{(\bar{x}_1 - \bar{x}_2) - (\mu_1 - \mu_2)}{\sqrt{\frac{s_1^2}{n_1} + \frac{s_2^2}{n_2}}} = \frac{(18.9 - 22.7) - (0)}{\sqrt{\frac{(6.7)^2}{73} + \frac{(7.6)^2}{102}}} \approx -3.496$$

Reject H_0. There is enough evidence to reject the claim.

4. (a) Standard normal distribution because $n > 30$.

(b) Standard normal distribution because $n_1 > 30$ and $n_2 > 30$ and the samples are independent.

5. $r \approx 0.989 \rightarrow$ Positive linear correlation; the number of acres harvested increases as the number of acres planted increases.

6. $\hat{y} = 1.071x - 268.253$

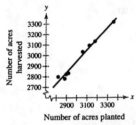

7. $r^2 = 0.978 \rightarrow 97.8\%$ of the variation in y is explained by the model.

8. $s_e = 31.46$

The standard deviation of the number of acres harvested for a specified number of acres planted is 31.46 thousand acres.

9. $\hat{y} = 1.071x - 268.253$

$\phantom{\hat{y}} = 1.071(3100) - 268.253$

$\phantom{\hat{y}} = 3051.847$

$$E = t_c s_e \sqrt{1 + \frac{1}{n} + \frac{n(x - \bar{x})^2}{n\Sigma x^2 - (\Sigma x)^2}}$$

$$= 2.447(31.46)\sqrt{1 + \frac{1}{8} + \frac{8(3100 - 3028.3)^2}{8(73,592,588) - 24,226^2}}$$

≈ 82.459

$\hat{y} - E < y < \hat{y} + E$

$3051.847 - 82.459 < y < 3051.847 + 82.459$

$2969.388 < y < 3134.306$

10. $H_0: \rho = 0$ and $H_a: \rho \neq 0$

Critical value is ± 3.707

$$t = \frac{r}{\sqrt{\frac{1 - r^2}{n - 2}}} = \frac{0.989}{\sqrt{\frac{1 - (0.989)^2}{8 - 2}}} = 16.378$$

Reject H_0. There is enough evidence to conclude that a significant correlation exists.

Chi-Square Tests and the *F*-Distribution

10.1 GOODNESS OF FIT

10.1 Try It Yourself Solutions

1a.

Music	%of Listeners	Expected Frequency
Classical	4%	12
Country	36%	108
Gospel	11%	33
Oldies	2%	6
Pop	18%	54
Rock	29%	87

2a. Claimed Distribution:

Ages	Distribution
0–9	16%
10–19	20%
20–29	8%
30–39	14%
40–49	15%
50–59	12%
60–69	10%
70+	5%

H_0: Distribution of ages is as shown in table above.

H_a: Distribution of ages differs from the claimed distribution.

b. $\alpha = 0.05$ **c.** d.f. $= n - 1 = 7$

d. Critical value is $14.067 \rightarrow$ Reject H_0 if $\chi^2 > 14.067$

e.

Ages	Distribution	Observed	Expected	$\frac{(O - E)^2}{E}$
0-9	16%	76	64	2.250
10–19	20%	84	80	0.200
20–29	8%	30	32	0.125
30–39	14%	60	56	0.286
40–49	15%	54	60	0.600
50–59	12%	40	48	1.333
60–69	10%	42	40	0.100
70+	5%	14	20	1.800
				6.694

$\chi^2 = 6.694$

f. Fail to reject H_0.

g. There is not enough evidence to conclude that the distribution of ages differs from the claimed distribution.

3a. Claimed Distribution:

Response	Distribution
In favor of	44%
Against	27%
No Opinion	29%

H_0: Distribution of responses is as shown in table above.

H_a: Distribution of responses differs from the claimed distribution.

b. $\alpha = 0.01$ **c.** d.f. $= n - 1 = 2$ **d.** Critical value is $9.210 \rightarrow$ Reject H_0 if $\chi^2 > 9.210$

e.

Response	Distribution	Observed	Expected	$\frac{(O - E)^2}{E}$
In favor of	44%	100	88	1.636
Against	27%	48	54	0.667
No opinion	29%	52	58	0.621
				2.924

f. Fail to reject H_0.

g. There is not enough evidence to conclude that the distribution of responses differs from the claimed distribution.

10.1 EXERCISE SOLUTIONS

1. (a) Claimed distribution:

Response	Distribution
Home	70%
Work	17%
Commuting	8%
Other	5%

H_0: Distribution of responses is as shown in table above.

H_a: Distribution of responses differs from the claimed distribution.

(b) Critical value is $7.815 \rightarrow$ Reject H_0 if $\chi^2 > 7.815$

(c)

Response	Distribution	Observed	Expected	$\frac{(O - E)^2}{E}$
Home	70%	389	406.70	0.770
Work	17%	110	98.77	1.277
Commuting	8%	55	46.48	1.562
Other	5%	27	29.05	0.145
				3.754

$\chi^2 = 3.754$

(d) Fail to reject H_0. There is not enough evidence to conclude that the distribution of the responses differs from the claimed distribution.

3. (a) Claimed distribution:

Day	Distribution
Sunday	14.286%
Monday	14.286%
Tuesday	14.286%
Wednesday	14.286%
Thursday	14.286%
Friday	14.286%
Saturday	14.286%

H_0: The distribution of fatal bicycle accidents throughout the week is as shown in the table above.

H_a: The distribution of fatal bicycle accidents throughout the week differs from the claimed distribution.

(b) Critical value is $10.645 \rightarrow$ Reject H_0 if $\chi^2 > 10.645$

(c)

Day	Distribution	Observed	Expected	$\frac{(O - E)^2}{E}$
Sunday	14.286%	118	130.15	1.133
Monday	14.286%	119	130.15	0.954
Tuesday	14.286%	127	130.15	0.076
Wednesday	14.286%	137	130.15	0.361
Thursday	14.286%	129	130.15	0.010
Friday	14.286%	146	130.15	1.931
Saturday	14.286%	135	130.15	0.181
				4.648

$\chi^2 = 4.648$

(d) Fail to reject H_0. There is not enough evidence to conclude that the distribution of fatal bicycle accidents throughout the week differs from the claimed distribution.

5. (a) Claimed distribution:

Object struck	Distribution
Tree	28%
Embankment	10%
Utility pole	10%
Guardrail	9%
Ditch	7%
Curb	6%
Culvert	5%
Sign/Post/Fence	10%
Other	15%

H_0: Distribution of objects struck is as shown in table above.

H_a: Distribution of objects struck differs from the claimed distribution.

(b) Critical value is $= 20.090 \rightarrow$ Reject H_0 if $\chi^2 > 20.090$

(c)

Object Struck	Distribution	Observed	Expected	$\frac{(O - E)^2}{E}$
Tree	28%	179	193.48	1.084
Embankment	10%	100	69.10	13.818
Utility pole	10%	107	69.10	20.787
Guardrail	9%	57	62.19	0.433
Ditch	7%	36	48.37	3.163
Curb	6%	43	41.46	0.057
Culvert	5%	28	34.55	1.242
Sign/Post/Fence	10%	68	69.10	0.018
Other	15%	73	103.65	9.063
			691	49.665

$\chi^2 = 49.665$

(d) Reject H_0. There is enough evidence to conclude that the distribution of the objects struck differs from the claimed distribution.

7. (a) Claimed distribution:

Response	Distribution
Not a HS grad	33.333%
HS graduate	33.333%
College (1yr+)	33.333%

H_0: Distribution of the responses is as shown in table above.

H_a: Distribution of the responses differs from the claimed distribution.

(b) Critical value is $7.378 \rightarrow$ Reject H_0 if $\chi^2 > 7.378$

(c)

Response	Distribution	Observed	Expected	$\frac{(O - E)^2}{E}$
Not a HS grad	33.333%	37	33	0.485
HS graduate	33.333%	40	33	1.485
College (1yr+)	33.333%	22	33	3.667
			99	5.637

$\chi^2 = 5.637$

(d) Fail to reject H_0. There is not enough evidence to conclude that the distribution of the responses differs from the claimed distribution.

9. (a) Claimed distribution:

Cause	Distribution
Trans. Accidents	41%
Assaults	20%
Objects/equipment	15%
Falls	10%
Exposure	10%
Other	4%

H_0: Distribution of the causes is as shown in table above.

H_a: Distribution of the causes differs from the claimed distribution.

(b) Critical value is $11.071 \rightarrow$ Reject H_0 if $\chi^2 > 11.071$

(c)

Cause	Distribution	Observed	Expected	$\frac{(O - E)^2}{E}$
Trans. Accidents	41%	2500	2554.71	1.172
Assaults	20%	1300	1246.20	2.323
Objects/equipment	15%	985	934.65	2.712
Falls	10%	620	623.10	0.015
Exposure	10%	602	623.10	0.715
Other	4%	224	249.24	2.556
		6231		9.493

$\chi^2 = 9.493$

(d) Fail to reject H_0. There is not enough evidence to conclude that the distributions of the causes differs from the claimed distribution.

11. (a) Frequency distribution: $\mu = 69.435$; $\sigma \approx 8.337$

Lower Boundary	Upper Boundary	Lower z-score	Upper z-score	Area
49.5	58.5	−2.39	−1.31	0.0867
58.5	67.5	−1.31	−0.23	0.3139
67.5	76.5	−0.23	0.85	0.3933
76.5	85.5	0.85	1.93	0.1709
85.5	94.5	1.93	3.01	0.0255

Class Boundaries	Distribution	Frequency	Expected	$\frac{(O-E)^2}{E}$
49.5–58.5	8.67%	19	17	0.235
58.5–67.5	31.39%	61	63	0.063
67.5–76.5	39.33%	82	79	0.114
76.5–85.5	17.09%	34	34	0
85.5–94.5	2.55%	4	5	0.2
		200		0.612

H_0: Variable has a normal distribution

H_a: Variable does not have a normal distribution

(b) Critical value is 13.277

(c) $\chi^2 = 0.612$

(d) Fail to reject H_0. There is not enough evidence to conclude that the variable does not have a normal distribution.

10.2 INDEPENDENCE

10.2 Try It Yourself Solutions

1ab.

	Hotel	Leg Room	Rental Size	Other	Total
Business	36	108	14	22	180
Leisure	38	54	14	14	120
Total	74	162	28	36	300

c. $n = 300$

d.

	Hotel	Leg Room	Rental Size	Other
Business	44.4	97.2	16.8	21.6
Leisure	29.6	64.8	11.2	14.4

2a. H_0: Travel concern is independent of travel purpose.

H_a: Travel concern is dependent on travel purpose. (claim)

b. $\alpha = 0.01$ **c.** $(r-1)(c-1) = 3$ **d.** Critical value is 11.345 → Reject H_0 if $\chi^2 > 11.345$

e. $\chi^2 = \sum \frac{(O-E)^2}{E} \approx 8.158$ **f.** Fail to reject H_0.

g. There is not enough evidence to conclude that travel concern is dependent on travel purpose.

3a. Critical value is $9.488 \rightarrow$ Reject H_0 if $\chi^2 > 9.488$ **b.** Enter the data.   **c.** $\chi^2 \approx 65.619$

d. Reject H_0.   **e.** Yes.

10.2 EXERCISE SOLUTIONS

1. (a) H_0: Skill level in a subject is independent of location.

H_a: Skill level in a subject is dependent on location.

(b) d.f. $= (r - 1)(c - 1) = 2$

Critical value is $9.210 \rightarrow$ Reject H_0 if $\chi^2 > 9.210$

(c) $\chi^2 \approx 0.297$

(d) Fail to reject H_0. There is not enough evidence to conclude that skill level in a subject is dependent on location.

3. (a) H_0: Adults' ratings are independent of the type of school.

H_a: Adults' ratings are dependent on the type of school.

(b) d.f. $= (r - 1)(c - 1) = 3$

Critical value is $7.815 \rightarrow$ Reject H_0 if $\chi^2 > 7.815$

(c) $\chi^2 \approx 148.389$

(d) Reject H_0. There is enough evidence to conclude that adults' ratings are dependent on the type of school.

5. (a) H_0: Results are independent of the type of treatment.

H_a: Results are dependent on the type of treatment.

(b) d.f. $= (r - 1)(c - 1) = 1$

Critical value is $2.706 \rightarrow$ Reject H_0 if $\chi^2 > 2.706$

(c) $\chi^2 \approx 5.106$

(d) Reject H_0. There is not enough evidence to conclude that results are dependent on the type of treatment.

7. (a) H_0: Reasons are independent of the type of worker.

H_a: Reasons are dependent on the type of worker.

(b) d.f. $= (r - 1)(c - 1) = 2$

Critical value is $9.210 \rightarrow$ Reject H_0 if $\chi^2 > 9.210$

(c) $\chi^2 \approx 7.326$

(d) Fail to reject H_0. There is not enough evidence to conclude that the reasons are dependent on the type of worker.

9. (a) H_0: Type of crash is independent of the type of vehicle.

H_a: Type of crash is dependent on the type of vehicle.

(b) d.f. $= (r - 1)(c - 1) = 2$

Critical value is $5.991 \rightarrow$ Reject H_0 if $\chi^2 > 5.991$

(c) $\chi^2 \approx 106.390$

(d) Reject H_0. There is enough evidence to conclude that type of crash is dependent on the type of vehicle.

11. H_0: The proportions are equal.

H_a: At least one of the proportions is different from the others.

d.f. $= (r - 1)(c - 1) = 7$

Critical value is 14.067 $\rightarrow$ Reject H_0 if $\chi^2 > 14.067$

$\chi^2 \approx 3.853$

Fail to reject H_0. There is not enough evidence to conclude that at least one of the proportions is different from the others.

10.3 COMPARING TWO VARIANCES

10.3 Try It Yourself Solutions

1a. $\alpha = 0.01$ **b.** d.f.$_N = 3$ **c.** d.f.$_D = 15$ **d.** $F = 5.42$

2a. $\alpha = 0.01$ **b.** d.f.$_N = 2$ **c.** d.f.$_D = 5$ **d.** $F = 13.27$

3a. H_0: $\sigma_1^2 \leq \sigma_2^2$ and H_a: $\sigma_1^2 > \sigma_2^2$ (claim)

b. $\alpha = 0.01$

c. d.f.$_N = n_1 - 1 = 24$

d.f.$_D = n_2 - 1 = 19$

d. Critical value is 2.92 $\rightarrow$ Reject H_0 if $F > 2.92$

e. $F = \dfrac{s_1^2}{s_2^2} = \dfrac{180}{56} \approx 3.214$

f. Reject H_0.

g. There is enough evidence to support the claim.

4a. H_0: $\sigma_1 = \sigma_2$ (claim) and H_a: $\sigma_1 \neq \sigma_2$

b. $\alpha = 0.01$

c. d.f.$_N = n_1 - 1 = 15$

d.f.$_D = n_2 - 1 = 21$

d. Critical value is 3.43 $\rightarrow$ Reject H_0 if $F > 3.43$

e. $F = \dfrac{s_1^2}{s_2^2} = \dfrac{(0.95)^2}{(0.78)^2} \approx 1.483$

f. Fail to reject H_0.

g. There is not enough evidence to reject the claim.

10.3 EXERCISE SOLUTIONS

1. Specify the level of significance α. Determine the degrees of freedom for the numerator and denominator. Use Table 7 to find the critical value F.

3. $F = 2.93$

5. $F = 5.32$

7. $H_0: \sigma_1^2 \leq \sigma_2^2$ and $H_a: \sigma_1^2 > \sigma_2^2$ (claim)

   d.f.$_N = 4$

   d.f.$_D = 5$

   Critical value is 3.52 $\rightarrow$ Reject H_0 if $F > 3.52$

   $F = \dfrac{s_1^2}{s_2^2} = \dfrac{773}{765} \approx 1.010$

   Fail to reject H_0. There is not enough evidence to support the claim.

9. $H_0: \sigma_1^2 \leq \sigma_2^2$ (claim) and $H_a: \sigma_1^2 > \sigma_2^2$

   d.f.$_N = 10$

   d.f.$_D = 9$

   Critical value is 5.26 $\rightarrow$ Reject H_0 if $F > 5.26$

   $F = \dfrac{s_1^2}{s_2^2} = \dfrac{842}{836} \approx 1.007$

   Fail to reject H_0. There is not enough evidence to reject the claim.

11. (a) Population 1: Company B

    Population 2: Company A

    $H_0: \sigma_1^2 \leq \sigma_2^2$ and $H_a: \sigma_1^2 > \sigma_2^2$ (claim)

    (b) d.f.$_N = 22$

    d.f.$_D = 19$

    Critical value is 2.13 $\rightarrow$ Reject H_0 if $F > 2.13$

    (c) $F = \dfrac{s_1^2}{s_2^2} = \dfrac{2.8}{2.6} \approx 1.077$

    (d) Fail to reject H_0. There is not enough evidence to support the claim.

13. (a) $H_0: \sigma_1^2 = \sigma_2^2$ (claim) and $H_a: \sigma_1^2 \neq \sigma_2^2$

    (b) d.f.$_N = 11$

    d.f.$_D = 13$

    Critical value is 2.63 $\rightarrow$ Reject H_0 if $F > 2.63$

    (c) $F = \dfrac{s_1^2}{s_2^2} = \dfrac{(27.7)^2}{(26.1)^2} \approx 1.126$

    (d) Fail to reject H_0. There is not enough evidence to reject the claim.

15. (a) $H_0: \sigma_1^2 \leq \sigma_2^2$ and $H_a: \sigma_1^2 > \sigma_2^2$ (claim)

    (b) d.f.$_N = 24$

    d.f.$_D = 20$

    Critical value is 1.77 $\rightarrow$ Reject H_0 if $F > 1.77$

    (c) $F = \dfrac{s_1^2}{s_2^2} = \dfrac{(0.7)^2}{(0.5)^2} \approx 1.96$

    (d) Reject H_0. There is enough evidence to support the claim.

17. (a) Population 1: California

    Population 2: New York

    $H_0: \sigma_1^2 \leq \sigma_2^2$ and $H_a: \sigma_1^2 > \sigma_2^2$ (claim)

(b) $\text{d.f.}_N = 15$
$\text{d.f.}_D = 16$
Critical value is $2.35 \rightarrow$ Reject H_0 if $F > 2.35$

(c) $F = \dfrac{s_1^2}{s_2^2} = \dfrac{(23,900)^2}{(18,800)^2} \approx 1.616$

(d) Fail to reject H_0. There is not enough evidence to support the claim.

19. Right-tailed: $F = 8.94$

Left-tailed:
(1) $\text{d.f.}_N = 3$ and $\text{d.f.}_D = 6$

(2) $F = 4.76$

(3) Critical value is $\dfrac{1}{F} = \dfrac{1}{4.76} \approx 0.210$

21. $\dfrac{s_1^2}{s_2^2}F_L < \dfrac{\sigma_1^2}{\sigma_2^2} < \dfrac{s_1^2}{s_2^2}F_R \rightarrow \dfrac{9.61}{8.41}0.32 < \dfrac{\sigma_1^2}{\sigma_2^2} < \dfrac{9.61}{8.41}3.36 \rightarrow 0.366 < \dfrac{\sigma_1^2}{\sigma_2^2} < 3.839$

10.4 ANALYSIS OF VARIANCE

10.4 Try It Yourself Solutions

1a. $H_0: \mu_1 = \mu_2 = \mu_3 = \mu_4$
H_a: At least one mean is different from the others.

b. $\alpha = 0.05$

c. $\text{d.f.}_N = 3$
$\text{d.f.}_D = 14$

d. Critical value is $3.34 \rightarrow$ Reject H_0 if $F > 3.34$

e.

Variation	Sum of Squares	Degrees of Freedom	Mean Squares	F
Between	549.8	3	183.3	4.22
Within	608.0	14	43.4	

$F \approx 4.22$

f. Reject H_0.

g. There is enough evidence to conclude that at least one mean is different from the others.

2a. Enter the data.

b. $H_0: \mu_1 = \mu_2 = \mu_3 = \mu_4$

H_a: At least one mean is different from the others.

Variation	Sum of Squares	Degrees of Freedom	Mean Squares	F
Between	0.584	3	0.195	1.34
Within	4.360	30	0.145	

$F = 1.34 \rightarrow P\text{-value} = 0.280$

c. Fail to reject H_0.

d. There is not enough evidence to conclude that at least one mean is different from the others.

10.4 EXERCISE SOLUTIONS

1. (a) H_0: $\mu_1 = \mu_2 = \mu_3$

H_a: At least one mean is different from the others. (claim)

(b) d.f.$_N = k - 1 = 2$

d.f.$_D = N - k = 27$

Critical value is 3.35 → Reject H_0 if $F > 3.35$

(c)

Variation	Sum of Squares	Degrees of Freedom	Mean Squares	F
Between	1.630	2	0.815	1.26
Within	17.469	27	0.647	

$F = 1.26$

(d) Fail to reject H_0. There is not enough evidence to support the claim.

3. (a) H_0: $\mu_1 = \mu_2 = \mu_3$ (claim)

H_a: At least one mean is different from the others.

(b) d.f.$_N = k - 1 = 2$

d.f.$_D = N - k = 12$

Critical value is 2.81 → Reject H_0 if $F > 2.81$

(c)

Variation	Sum of Squares	Degrees of Freedom	Mean Squares	F
Between	302.6	2	151.3	1.77
Within	1024.3	12	85.4	

$F = 1.77$

(d) Fail to reject H_0. There is not enough evidence to reject the claim.

5. (a) H_0: $\mu_1 = \mu_2 = \mu_3 = \mu_4$ (claim)

H_a: At least one mean is different from the others.

(b) d.f.$_N = k - 1 = 3$

d.f.$_D = N - k = 33$

Critical value is 4.44 → Reject H_0 if $F > 4.44$

(c)

Variation	Sum of Squares	Degrees of Freedom	Mean Squares	F
Between	36.50	3	12.17	5.21
Within	77.06	33	2.34	

$F = 5.21$

(d) Reject H_0. There is not enough evidence to reject the claim.

7. (a) H_0: $\mu_1 = \mu_2 = \mu_3 = \mu_4$ (claim)

H_a: At least one mean is different from the others.

(b) $\text{d.f.}_N = k - 1 = 3$

$\text{d.f.}_D = N - k = 43$

Critical value is 2.22 → Reject H_0 if $F > 2.22$

(c)

Variation	Sum of Squares	Degrees of Freedom	Mean Squares	F
Between	15,532	3	5177	3.04
Within	73,350	43	1706	

$F = 3.04$

(d) Reject H_0. There is not enough evidence to reject the claim.

9. (a) $H_0: \mu_1 = \mu_2 = \mu_3 = \mu_4$

H_a: At least one mean is different from the others. (claim)

(b) $\text{d.f.}_N = k - 1 = 3$

$\text{d.f.}_D = N - k = 36$

Critical value is 4.38 → Reject H_0 if $F > 4.38$

(c)

Variation	Sum of Squares	Degrees of Freedom	Mean Squares	F
Between	61,131	3	20,377	8.46
Within	86,713	36	2409	

$F = 8.46$

(d) Reject H_0. There is enough evidence to support the claim.

11.

	Mean	Size
Pop 1	7.400	10
Pop 2	5.900	10
Pop 3	4.875	8
Pop 4	5.111	9

$SS_w = 77.06$

$\Sigma(n_i - 1) = N - k = 33$

Critical value is $4.44 \rightarrow CV_{\text{Scheffe}} = 4.44(4 - 1) = 13.320$

$$\frac{(\bar{x}_1 - \bar{x}_2)^2}{\dfrac{SS_w}{\Sigma(n_i - 1)}\left[\dfrac{1}{n_1} + \dfrac{1}{n_2}\right]} \approx 4.818 \rightarrow \text{No difference}$$

$$\frac{(\bar{x}_1 - \bar{x}_2)^2}{\dfrac{SS_w}{\Sigma(n_i - 1)}\left[\dfrac{1}{n_1} + \dfrac{1}{n_2}\right]} \approx 12.135 \rightarrow \text{No difference}$$

$$\frac{(\bar{x}_1 - \bar{x}_2)^2}{\dfrac{SS_w}{\Sigma(n_i - 1)}\left[\dfrac{1}{n_1} + \dfrac{1}{n_2}\right]} \approx 10.628 \rightarrow \text{No difference}$$

$$\frac{(\bar{x}_1 - \bar{x}_2)^2}{\dfrac{SS_w}{\Sigma(n_i - 1)}\left[\dfrac{1}{n_1} + \dfrac{1}{n_2}\right]} \approx 2.000 \rightarrow \text{No difference}$$

$$\frac{(\bar{x}_1 - \bar{x}_2)^2}{\frac{SS_w}{\Sigma(n_i - 1)}\left[\frac{1}{n_1} + \frac{1}{n_2}\right]} \approx 1.263 \rightarrow \text{No difference}$$

$$\frac{(\bar{x}_1 - \bar{x}_2)^2}{\frac{SS_w}{\Sigma(n_i - 1)}\left[\frac{1}{n_1} + \frac{1}{n_2}\right]} \approx 0.101 \rightarrow \text{No difference}$$

13.

	Mean	Size
Pop 1	139.35	12
Pop 2	93.05	11
Pop 3	107.72	12
Pop 4	130.75	12

$SS_w = 73,350$

$\Sigma(n_i - 1) = N - k = 43$

Critical value is $2.22 \rightarrow CV_{\text{Scheffe}} = 2.22(4 - 1) = 6.660$

$$\frac{(\bar{x}_1 - \bar{x}_2)^2}{\frac{SS_w}{\Sigma(n_i - 1)}\left[\frac{1}{n_1} + \frac{1}{n_2}\right]} \approx 7.212 \rightarrow \text{Significant difference}$$

$$\frac{(\bar{x}_1 - \bar{x}_2)^2}{\frac{SS_w}{\Sigma(n_i - 1)}\left[\frac{1}{n_1} + \frac{1}{n_2}\right]} \approx 3.519 \rightarrow \text{No difference}$$

$$\frac{(\bar{x}_1 - \bar{x}_2)^2}{\frac{SS_w}{\Sigma(n_i - 1)}\left[\frac{1}{n_1} + \frac{1}{n_2}\right]} \approx 0.260 \rightarrow \text{No difference}$$

$$\frac{(\bar{x}_1 - \bar{x}_2)^2}{\frac{SS_w}{\Sigma(n_i - 1)}\left[\frac{1}{n_1} + \frac{1}{n_2}\right]} \approx 0.724 \rightarrow \text{No difference}$$

$$\frac{(\bar{x}_1 - \bar{x}_2)^2}{\frac{SS_w}{\Sigma(n_i - 1)}\left[\frac{1}{n_1} + \frac{1}{n_2}\right]} \approx 4.782 \rightarrow \text{No difference}$$

$$\frac{(\bar{x}_1 - \bar{x}_2)^2}{\frac{SS_w}{\Sigma(n_i - 1)}\left[\frac{1}{n_1} + \frac{1}{n_2}\right]} \approx 1.866 \rightarrow \text{No difference}$$

CHAPTER 10 REVIEW EXERCISE SOLUTIONS

1. Claimed distribution:

Category	Distribution
New Patients	25%
Old/New	25%
Old/Recurring	50%

H_0: Distribution of office visits is as shown in table above.

H_a: Distribution of office visits differs from the claimed distribution.

Critical value is 5.991

Category	Distribution	Observed	Expected	$\frac{(O - E)^2}{E}$
New Patients	25%	97	174	34.075
Old/New	25%	142	174	5.885
Old/Recurring	50%	457	348	34.141
		696		74.101

$\chi^2 = 74.101$

Reject H_0. There is enough evidence to conclude that the distribution of office visits is not as claimed.

3. (a) Expected frequencies:

	HS-Did not Complete	HS Complete	College 1-3 years	College 4+ years	Total
25-44	17,303.105	20,295.862	10,260.752	12,370.281	60,230
45+	20,271.895	23,778.138	12,021.248	14,492.719	70,564
Total	37,575	44,074	22,282	26,863	130,794

(b) H_0: Education is independent of age.

H_a: Education is dependent on age.

Critical value is 6.251

$$\chi^2 = \Sigma\frac{(O - E)^2}{E} = 11{,}852.71$$

Reject H_0

(c) There is enough evidence to conclude that education is dependent on age.

5. $F = 2.295$

7. $F = 2.39$

9. H_0: $\sigma_1^2 \geq \sigma_2^2$ (claim) and H_a: $\sigma_1^2 < \sigma_2^2$ (left-tailed test)

d.f.$_N$ = 15

d.f.$_D$ = 20

Critical value is 3.09 → Reject H_0 if $F > 3.09$

$F = \frac{s_1^2}{s_2^2} = \frac{653}{270} \approx 2.419$

Fail to reject H_0. There is not enough evidence to reject the claim.

11. H_0: $\sigma_1^2 \leq \sigma_2^2$ and H_a: $\sigma_1^2 > \sigma_2^2$ (claim)

d.f.$_N$ = 20

d.f.$_D$ = 15

Critical value is 1.92 → Reject H_0 if $F > 1.92$

$F = \frac{s_1^2}{s_2^2} = \frac{(0.76)^2}{(0.58)^2} \approx 1.717$

Fail to reject H_0. There is not enough evidence to support the claim.

13. Population 1: Male → $s_1^2 = 18{,}486.26$

Population 2: Female → $s_2^2 = 12{,}102.78$

H_0: $\sigma_1^2 = \sigma_2^2$ and H_a: $\sigma_1^2 \neq \sigma_2^2$ (claim)

d.f.$_N$ = 13

d.f.$_D$ = 8

Critical value is $6.94 \rightarrow$ Reject H_0 if $F > 6.94$

$$F = \frac{s_1^2}{s_2^2} = \frac{18{,}486.26}{12{,}102.78} \approx 1.527$$

Fail to reject H_0. There is not enough evidence to support the claim.

15. H_0: $\mu_1 = \mu_2 = \mu_3 = \mu_4$

   H_a: At least one mean is different from the others. (claim)

   d.f.$_N = k - 1 = 3$

   d.f.$_D = N - k = 28$

   Critical value is $2.29 \rightarrow$ Reject H_0 if $F > 2.29$

Variation	Sum of Squares	Degrees of Freedom	Mean Squares	F
Between	2,318,032	3	772,677	6.60
Within	3,277,311	28	117,047	

   Reject H_0. There is enough evidence to support the claim.

CHAPTER 10 QUIZ SOLUTIONS

1. (a) Population 1: San Jose $\rightarrow s_1^2 \approx 429.984$

   Population 2: Dallas $\rightarrow s_2^2 \approx 112.779$

   H_0: $\sigma_1^2 = \sigma_2^2$ and H_a: $\sigma_1^2 \neq \sigma_2^2$ (claim)

   (b) $\alpha = 0.05$

   (cd) d.f.$_N = 12$

   d.f.$_D = 17$

   Critical value is $2.82 \rightarrow$ Reject H_0 if $F > 2.82$

   (e) $F = \dfrac{s_1^2}{s_2^2} = \dfrac{429.984}{112.779} \approx 3.813$

   (fg) Reject H_0. There is enough evidence to support the claim.

2. (a) H_0: $\mu_1 = \mu_2 = \mu_3$ (claim)

   H_a: At least one mean is different from the others.

   (b) $\alpha = 0.10$

   (cd) d.f.$_N = k - 1 = 2$

   d.f.$_D = N - k = 44$

   Critical value is $2.43 \rightarrow$ Reject H_0 if $F > 2.43$

Variation	Sum of Squares	Degrees of Freedom	Mean Squares	F
Between	2676	2	1338	7.39
Within	7962	44	181	

   (e) $F = 7.39$

   (fg) Reject H_0. There is enough evidence to reject the claim.

3. (a) Claimed distribution:

Education	25 & Over
Not a HS graduate	18.3%
HS graduate	33.6%
Some college, no degree	17.3%
Associate Degree	7.2%
Bachelor Degree	15.8%
Advanced Degree	7.8%

H_0: Distribution of educational achievement is as shown in table above.

H_a: Distribution of educational achievement differs from the claimed distribution.

(b) $\alpha = 0.01$

(cd) Critical value is $15.086 \rightarrow$ Reject H_0 if $\chi^2 > 15.086$

(e)

Education	25 & Over	Observed	Expected	$\frac{(O-E)^2}{E}$
Not a HS graduate	18.3%	37	55.083	5.936
HS graduate	33.6%	102	101.136	0.007
Some college, no degree	17.3%	59	53.073	0.921
Associate Degree	7.2%	27	21.672	1.310
Bachelor Degree	15.8%	51	47.558	0.249
Advanced Degree	7.8%	25	23.478	0.099
		301		8.522

$\chi^2 = 8.522$

(fg) Fail to reject H_0. There is not enough evidence to conclude that the distribution of educational achievement differs from the claimed distribution.

4. (a) Claimed distribution:

Education	25 & Over
Not a HS graduate	18.3%
HS graduate	33.6%
Some college, no degree	17.3%
Associate Degree	7.2%
Bachelor Degree	15.8%
Advanced Degree	7.8%

H_0: Distribution of educational achievement is as shown in table above.

H_a: Distribution of educational achievement differs from the claimed distribution.

(b) $\alpha = 0.05$

(c) Critical value is 11.071

(d) Reject H_0 if $\chi^2 > 11.071$

(e)

Education	25 & Over	Observed	Expected	$\frac{(O - E)^2}{E}$
Not a HS graduate	18.3%	124	74.481	32.290
HS graduate	33.6%	148	136.752	0.925
Some college, no degree	17.3%	61	70.411	1.258
Associate Degree	7.2%	15	29.304	6.982
Bachelor Degree	15.8%	37	64.306	11.595
Advanced Degree	7.8%	22	31.746	2.992
		407		56.042

$\chi^2 = 56.042$

(f) Reject H_0.

(g) There is enough evidence to conclude that the distribution of educational achievement differs from the claimed distribution.

Nonparametric Tests

11.1 THE SIGN TEST

11.1 Try It Yourself Solutions

1a. H_0: median ≤ 2500 and H_a: median > 2500 (claim)

b. $\alpha = 0.025$

c. $n = 22$

d. The critical value is 5.

e. $x = 10$

f. Fail to reject H_0.

g. There is not enough evidence to support the claim.

2a. H_0: median $= 134{,}500$ (claim) and H_a: median $\neq 134{,}500$

b. $\alpha = 0.10$

c. $n = 81$

d. The critical values are ± 1.645.

e. $x = 30$
$$z = \frac{(x + 0.5) - 0.5(n)}{\frac{\sqrt{n}}{2}} = \frac{(30 + 0.5) - 0.5(81)}{\frac{\sqrt{81}}{2}} = -2.22$$

f. Reject H_0.

g. There is enough evidence to reject the claim.

3a. H_0: The number of colds will not decrease.
H_a: The number of colds will decrease. (claim)

b. $\alpha = 0.05$

c. $n = 11$

d. The critical value is 2.

e. $x = 2$

f. Reject H_0.

g. There is enough evidence to support the claim.

11.1 EXERCISE SOLUTIONS

1. A nonparametric test is a hypothesis test that does not require any specific conditions concerning the shape of populations or the value of any population parameters.

A nonparametric test is usually easier to perform than its corresponding parametric test, but the nonparametric test is usually less efficient.

3. (a) H_0: median ≤ 200 and H_a: median > 200 (claim)

   (b) Critical value is 1.

   (c) $x = 5$

   (d) Fail to reject H_0. There is not enough evidence to support the claim.

5. (a) H_0: median $\leq 140{,}000$ (claim)

   H_a: median $> 140{,}000$

   (b) Critical value is 1.

   (c) $x = 3$

   (d) Fail to reject H_0. There is not enough evidence to reject the claim.

7. (a) H_0: median ≥ 1500 (claim) and H_a: median < 1500

   (b) Critical value is 2.055

   (c) $x = 36$

   $$z = \frac{(x + 0.5) - 0.5(n)}{\dfrac{\sqrt{n}}{2}} = \frac{(36 + 0.5) - 0.5(104)}{\dfrac{\sqrt{104}}{2}} \approx -3.040$$

   (d) Reject H_0. There is enough evidence to reject the claim.

9. (a) H_0: median ≤ 36 and H_a: median > 36 (claim)

   (b) Critical value is 3.

   (c) $x = 8$

   (d) Fail to reject H_0. There is not enough evidence to support the claim.

11. (a) H_0: median $= 5$ (claim) and H_a: median $\neq 5$

   (b) Critical value is -1.96.

   (c) $x = 15$

   $$z = \frac{(x + 0.5) - 0.5(n)}{\dfrac{\sqrt{n}}{2}} = \frac{(15 + 0.5) - 0.5(47)}{\dfrac{\sqrt{47}}{2}} \approx -2.334$$

   (d) Reject H_0. There is enough evidence to reject the claim.

13. (a) H_0: median $= \$9.81$ (claim) and H_a: median $\neq \$9.81$

   (b) Critical value is -2.575.

   (c) $x = 16$

   $$z = \frac{(x + 0.5) - 0.5(n)}{\dfrac{\sqrt{n}}{2}} = \frac{(16 + 0.5) - 0.5(39)}{\dfrac{\sqrt{39}}{2}} \approx -0.961$$

   (d) Fail to reject H_0. There is not enough evidence to reject the claim.

15. (a) H_0: The headache hours have not decreased.

   H_a: The headache hours have decreased. (claim)

(b) Critical value is 1.

(c) $x = 3$

(d) Fail to reject H_0. There is not enough evidence to support the claim.

17. (a) H_0: The SAT scores have not improved.

   H_a: The SAT scores have improved. (claim)

   (b) Critical value is 2.

   (c) $x = 4$

   (d) Fail to reject H_0. There is not enough evidence to support the claim.

19. (a) H_0: The proportion of companies offering transportation is equal to the proportion of companies that do not. (claim)

   H_a: The proportion of companies offering transportation is not equal to the proportion of companies that do not.

   Critical value is 5.

   $\alpha = 0.05$

   $x = 8$

   (b) Fail to reject H_0. There is not enough evidence to reject the claim.

21. (a) H_0: median ≤ 418 (claim) and H_a: median > 418

   (b) Critical value is 2.33.

   (c) $x = 29$

   $$z = \frac{(x + 0.5) - 0.5(n)}{\frac{\sqrt{n}}{2}} = \frac{(29 + 0.5) - 0.5(47)}{\frac{\sqrt{47}}{2}} \approx 1.459$$

   (d) Fail to reject H_0. There is not enough evidence to reject the claim.

23. (a) H_0: median ≤ 24 and H_a: median > 24 (claim)

   (b) Critical value is 1.645.

   (c) $x = 38$

   $$z = \frac{(x + 0.5) - 0.5(n)}{\frac{\sqrt{n}}{2}} = \frac{(38 + 0.5) - 0.5(60)}{\frac{\sqrt{60}}{2}} \approx 1.936$$

   (d) Reject H_0. There is enough evidence to support the claim.

11.2 THE WILCOXON TESTS

11.2 Try It Yourself Solutions

1a. H_0: The water repellent is not effective.

   H_a: The water repellent is effective. (claim)

b. $\alpha = 0.01$

c. $n = 11$

d. Critical value is 5.

e. $w_s = 10.5$

f. Fail to reject H_0.

g. There is not enough evidence to support the claim.

2a. H_0: There is no difference in the claims paid by the companies.
  H_a: There is a difference in the claims paid by the companies. (claim)

b. $\alpha = 0.05$

c. The critical values are ± 1.96.

d. $n_1 = 12$ and $n_2 = 12$

e. $R = 120.5$

f. $\mu_R = \dfrac{n_1(n_1 + n_2 + 1)}{2} = 150$

  $\sigma_R = \sqrt{\dfrac{n_1 n_2(n_1 + n_2 + 1)}{12}} = 17.321$

  $z = \dfrac{R - \mu_R}{\sigma_R} = -1.703$

g. Fail to reject H_0.

h. There is not enough evidence to support the claim.

11.2 EXERCISE SOLUTIONS

1. (a) H_0: There is no reduction in systolic blood pressure. (claim)
  H_a: There is a reduction in systolic blood pressure.

  (b) Wilcoxon signed-rank test

  (c) Critical value is 6.

  (d) $w_s = 6$

  (e) Reject H_0. There is enough evidence to support the claim.

3. (a) H_0: There is no difference in the earnings.
  H_a: There is a difference in the earnings. (claim)

  (b) Wilcoxon rank sum test

  (c) The critical values are ± 1.96

  (d) $R = 55$

  $\mu_R = \dfrac{n_1(n_1 + n_2 + 1)}{2} = 110$

  $\sigma_R = \sqrt{\dfrac{n_1 n_2(n_1 + n_2 + 1)}{12}} = 14.201$

  $z = \dfrac{R - \mu_R}{\sigma_R} = -3.873$

  (e) Reject H_0. There is enough evidence to support the claim.

5. (a) H_0: There is not a difference in salaries.

H_a: There is a difference in salaries. (claim)

(b) Wilcoxon rank sum test

(c) The critical values are ± 1.96

(d) $R = 118.5$

$$\mu_R = \frac{n_1(n_1 + n_2 + 1)}{2} = 150$$

$$\sigma_R = \sqrt{\frac{n_1 n_2(n_1 + n_2 + 1)}{12}} = 17.321$$

$$z = \frac{R - \mu_R}{\sigma_R} = -1.819$$

(e) Fail to reject H_0. There is not enough evidence to support the claim.

7. H_0: The fuel additive does not improve gas mileage.

H_a: The fuel additive does improve gas mileage. (claim)

Critical value is 1.282

$w_s = 43.5$

$$z = \frac{w_s - \frac{n(n + 1)}{4}}{\sqrt{\frac{n(n + 1)(2n + 1)}{24}}} = \frac{43.5 - \frac{32(32 + 1)}{4}}{\sqrt{\frac{32(32 + 1)[(2)32 + 1]}{24}}} \approx -4.123$$

Note: $n = 32$ because one of the differences is zero and should be discarded.

Reject H_0. There is enough evidence to support the claim.

11.3 THE KRUSKAL-WALLIS TEST

11.3 Try It Yourself Solutions

1a. H_0: There is no difference in the salaries in the three states.

H_a: There is a difference in the salaries in the three states. (claim)

b. $\alpha = 0.10$

c. d.f. $= k - 1 = 2$

d. Critical value is 4.605 $\rightarrow$ Reject H_0 if $\chi^2 > 4.605$

e.

State	Salary	Rank
CT	25.57	1
CA	26.42	2
NJ	29.73	3
CT	29.86	4
CT	30.00	5
CA	33.68	6.5
CT	33.68	6.5
NJ	33.91	8
NJ	34.29	9
NJ	36.35	10
CA	36.55	11
CA	37.18	12
NJ	37.24	13
CT	37.39	14
CA	38.36	15
CA	40.31	16
CA	41.33	17
NJ	43.26	18
NJ	43.89	19
NJ	44.57	20
CT	45.04	21
CT	45.29	22
CA	46.55	23
NJ	46.72	24
CA	47.17	25
CA	48.46	26
NJ	50.16	27
CT	51.03	28
CT	57.07	29
CT	61.46	30

$R_1 = 153.5$ $R_2 = 160.5$ $R_1 = 151$

f. $H = \dfrac{12}{N(N+1)}\left(\dfrac{R_1^2}{n_1} + \dfrac{R_2^2}{n_2} + \dfrac{R_3^2}{n_3}\right) - 3(N+1)$

$= \dfrac{12}{30(30+1)}\left(\dfrac{(153.5)^2}{10} + \dfrac{(160.5)^2}{10} + \dfrac{(151)^2}{10}\right) - 3(30+1) = 0.063$

g. Fail to reject H_0

h. There is not enough evidence to support the claim.

11.3 EXERCISE SOLUTIONS

1. (a) H_0: There is no difference in the premiums.

 H_a: There is a difference in the premiums. (claim)

 (b) Critical value is 5.991.

 (c) $H = 14.05$

 (d) Reject H_0. There is enough evidence to support the claim.

3. (a) H_0: There is no difference in the salaries.

 H_a: There is a difference in the salaries. (claim)

 (b) Critical value is 6.251.

(c) $H = 6.46$

(d) Reject H_0. There is enough evidence to support the claim.

5. (a) H_0: There is no difference in the number of days spent in the hospital.

H_a: There is a difference in the number of days spent in the hospital. (claim)

(b)

Variation	Sum of squares	Degrees of freedom	Mean Squares	F
Between	23.96	3	7.99	1.14
Within	231.12	33	7.00	

For $\alpha = 0.01$, the critical value is about 4.45. Because $F = 1.14$ is less than the critical value, the decision is to fail to reject H_0. There is not enough evidence to support the claim. This is the same decision found in part (a) using the Kruskal-Wallis test.

11.4 RANK CORRELATION

11.4 Try It Yourself Solutions

1a. $H_0: \rho_s = 0$ and $H_a: \rho_s \neq 0$

b. $\alpha = 0.05$

c. Critical value is 0.738.

d.

Oat	Rank	Wheat	Rank	d	d^2
1.49	6	3.72	6	0	0
1.14	1	2.61	1	0	0
1.21	2	3	2	0	0
1.32	4	3.24	3	1	1
1.36	5	3.26	4	1	1
1.22	3	3.45	5	−2	4
1.67	7	4.55	8	−1	1
1.9	8	4.3	7	1	1
					8

$\Sigma d^2 = 8$

e. $r_s = 1 - \dfrac{6\Sigma d^2}{n(n^2 - 1)} = 0.905$

f. Reject H_0.

g. There is enough evidence to conclude that a significant correction exists.

11.4 EXERCISE SOLUTIONS

1. (a) $H_0: \rho_s = 0$ and $H_a: \rho_s \neq 0$ (claim)

(b) Critical value is 0.929.

(c) $\Sigma d^2 = 4$

$$r_s = 1 - \dfrac{6\Sigma d^2}{n(n^2 - 1)} \approx 0.929$$

(d) Reject H_0. There is enough evidence to support the claim.

3. (a) $H_0: \rho_s = 0$ and $H_a: \rho_s \neq 0$ (claim)

(b) Critical value is 0.497.

(c) $\Sigma d^2 = 123.5$

$$r_s = 1 - \frac{6\Sigma d^2}{n(n^2 - 1)} \approx 0.568$$

(d) Reject H_0. There is enough evidence to support the claim.

5. $H_0: \rho_s = 0$ and $H_a: \rho_s \neq 0$ (claim)

Critical value is 0.700.

$\Sigma d^2 = 124$

$$r_s = 1 - \frac{6\Sigma d^2}{n(n^2 - 1)} \approx -0.033$$

Fail to reject H_0. There is not enough evidence to support the claim.

7. $H_0: \rho_s = 0$ and $H_a: \rho_s \neq 0$ (claim)

Critical value is 0.700.

$\Sigma d^2 = 42$

$$r_s = 1 - \frac{6\Sigma d^2}{n(n^2 - 1)} \approx 0.650$$

Fail to reject H_0. There is not enough evidence to support the claim.

9. $H_0: \rho_s = 0$ and $H_a: \rho_s \neq 0$ (claim)

$$\text{Critical value} = \frac{\pm z}{\sqrt{n - 1}} = \frac{\pm 1.96}{\sqrt{34 - 1}} \approx \pm 0.341$$

$\Sigma d^2 = 7310.5$

$$r_s = 1 - \frac{6\Sigma d^2}{n(n^2 - 1)} \approx -0.117$$

Fail to reject H_0. There is not enough evidence to support the claim.

CHAPTER 11 REVIEW EXERCISE SOLUTIONS

1. (a) $H_0:$ median = \$13,500 (claim) $H_a:$ median $\neq$ \$13,500

(b) Critical value is 2.

(c) $x = 7$

(d) Fail to reject H_0. There is not enough evidence to reject the claim.

3. (a) $H_0:$ median ≤ 6 (claim)

  $H_a:$ median > 6

(b) Critical value is 1.282.

(c) $x = 44$

$$z = \frac{(x - 0.5) - 0.5(n)}{\frac{\sqrt{n}}{2}} = \frac{(44 - 0.5) - 0.5(70)}{\frac{\sqrt{70}}{2}} = 2.032$$

(d) Reject H_0. There is enough evidence to reject the claim.

5. (a) H_0: There is no reduction in systolic blood pressure.(claim)
H_a: There is a reduction in systolic blood pressure.

(b) Critical value is 2.

(c) $x = 3$

(d) Fail to reject H_0. There is not enough evidence to reject the claim.

7. (a) Dependent; Wilcoxon Signed Rank Test

(b) H_0: Producers are not under reporting the caloric content of their foods.
H_a: Producers are under reporting the caloric content of their foods. (claim)

(c) Critical value is 8.

(d) $w_s = 2$

(e) Reject H_0. There is enough evidence to support the claim.

9. (a) Independent; Wilcoxon Rank Sum Test

(b) H_0: There is no difference in the amount of time that it takes to earn a doctorate.
H_a: There is a difference in the amount of time that it takes to earn a doctorate. (claim)

(c) Critical values are ± 2.575

(d) $R = 95$

$$\mu_R = \frac{n_1(n_1 + n_2 + 1)}{2} = 150$$

$$\sigma_R = \sqrt{\frac{n_1 n_2 (n_1 + n_2 + 1)}{12}} = 17.321$$

$$z = \frac{R - \mu_R}{\sigma_R} = -3.175$$

(e) Reject H_0. There is enough evidence to support the claim.

11. (a) H_0: There is no difference in salaries between the fields of study. (claim)
H_a: There is a difference in salaries between the fields of study.

(b) Critical value is 5.991.

(c) $H \approx 22.98$

(d) Reject H_0. There is enough evidence to reject the claim.

13. (a) H_0: $\rho_s = 0$ and H_a: $\rho_s \neq 0$ (claim)

(b) Critical value is 0.881.

(c) $\Sigma d^2 = 120$

$$r_s = 1 - \frac{6\Sigma d^2}{n(n^2 - 1)} = -0.429$$

(d) Fail to reject H_0. There is not enough evidence to support the claim.

CHAPTER 11 QUIZ SOLUTIONS

1. (a) H_0: There is no difference in the salaries between genders.
H_a: There is a difference in the salaries between genders. (claim)

(b) Wilcoxon Ranked Sum Test

(c) Critical values are ± 1.645.

(d) $R = 66$

$$\mu_R = \frac{n_1(n_1 + n_2 + 1)}{2} = 85.500$$

$$\sigma_R = \sqrt{\frac{n_1 n_2 (n_1 + n_2 + 1)}{12}} = 11.325$$

$$z = \frac{R - \mu_R}{\sigma_R} = -1.722$$

(e) Reject H_0. There is enough evidence to support the claim.

2. (a) H_0: median = 28 (claim) and H_a: median $\neq$ 28

(b) Sign Test

(c) Critical value is 6.

(d) $x = 10$

(e) Fail to reject H_0. There is not enough evidence to reject the claim.

3. (a) H_0: $\rho_s = 0$ and H_a: $\rho_s \neq 0$ (claim)

(b) Spearman Rank Correlation Coefficient Test

(c) Critical value is 0.881.

(d) $\Sigma d^2 = 76$

$$r_s = 1 - \frac{6\Sigma d^2}{n(n^2 - 1)} = 0.095$$

(e) Fail to reject H_0. There is not enough evidence to support the claim.

4. (a) H_0: There is no difference in the annual premiums between the states.
H_a: There is a difference in the annual premiums between the states. (claim)

(b) Kruskal-Wallis Test

(c) Critical value is 5.991.

(d) $H = 1.43$

(e) Fail to reject H_0. There is not enough evidence to support the claim.

CUMULATIVE TEST SOLUTIONS FOR CHAPTERS 10–11

1. (Sign Test)
H_0: median $\geq$ \$40,000
H_a: median < \$40,000 (claim)
Critical value is 1.
$x = 3$
Fail to reject H_0. There is not enough evidence to support the claim.

2. (Sign Test)

H_0: median $\leq$ \$29,000

H_a: median > \$29,000 (claim)

Critical value is 1.

$x = 4$

Fail to reject H_0. There is not enough evidence to support the claim.

3. (Wilcoxon Rank Sum Test)

H_0: There is no difference in the household incomes between the states. (claim)

H_a: There is a difference in the household incomes between the states.

Critical value is 2.575.

$R = 67$

$$\mu_R = \frac{n_1(n_1 + n_2 + 1)}{2} = 105$$

$$\sigma_R = \sqrt{\frac{n_1 n_2 (n_1 + n_2 + 1)}{12}} = 13.229$$

$$z = \frac{R - \mu_R}{\sigma_R} = -2.873$$

Reject H_0. There is enough evidence to reject the claim.

4. H_0: $\sigma_1^2 = \sigma_2^2$ claim and H_a: $\sigma_1^2 \neq \sigma_2^2$

d.f.$_N$ = 9

d.f.$_D$ = 9

Critical value is 5.35

$$F = \frac{s_1^2}{s_2^2} = \frac{99.034}{84.065} = 1.178$$

Fail to reject H_0. There is not enough evidence to reject the claim.

5. H_0: The distribution of Kentucky household incomes is the same as the U.S. distribution. (claim)

H_a: The distribution of Kentucky household incomes is not the same as the U.S. distribution.

Critical value is 10.645

$\chi^2 = 193.996$

Reject H_0. There is enough evidence to reject the claim.

6. H_0: The distribution of Utah household incomes is the same as the U.S. distribution. (claim)

H_a: The distribution of Utah household incomes is not the same as the U.S. distribution.

Critical value is 10.645

$\chi^2 = 39.240$

Reject H_0. There is enough evidence to reject the claim.

Appendix

Try It Yourself Solutions

3a. 0.4857

  b. $z = -2.17$ and $z = 2.17$

4a.

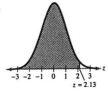

  b. 0.4834

  c. Area $= 0.5 + 0.4834 = 0.9834$

5a.

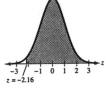

  b. 0.4846

  c. Area $= 0.5 + 0.4846 = 0.9846$

6a.

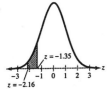

  b. $z = -2.16$: Area $= 0.4846$

    $z = -1.35$: Area $= 0.4115$

  c. Area $= 0.4846 - 0.4115 = 0.0731$